W9-AFT-067

ADVANCED MANUFACTURING TECHNOLOGY

SOCIETY OF MANUFACTURING ENGINEERS & DELMAR PUBLISHERS INC.

A PARTNERSHIP IN EDUCATIONAL EXCELLENCE

SME and DELMAR have proudly joined forces to form a partnership dedicated to educational excellence. We believe that quality manufacturing education is the key to keeping America competitive in the years ahead.

The Society of Manufacturing Engineers is an international technical society dedicated to advancing scientific knowledge in the field of manufacturing. SME has more than 80,000 members in 70 countries and serves as a forum for engineers and managers to share ideas, information, and accomplishments.

To be successful, today's engineers and technicians must keep pace with the torrent of information that appears each day. To meet this need, SME provides, in addition to the publication of books, many opportunities in continuing education for its members. These opportunities include: monthly meetings through five associations and more than 300 chapters; educational programs including seminars, clinics, and videotapes, as well as conferences and expositions.

Today's manufacturing technology students represent our future. Our goal is to provide these students with the finest manufacturing technology educational products. By pooling our many resources, SME and DELMAR are going to help teachers get the job done.

Together SME and DELMAR will provide outstanding educational materials to prepare students to enter the real world of manufacturing.

Thomas J. Drozda
Director of Publications
Society of Manufacturing Engineers

Gregory C. Spatz
President
Delmar Publishers Inc.

JAN 2 4 2000

TS
183
.G64
1990

ADVANCED MANUFACTURING TECHNOLOGY

DAVID L. GOETSCH

DELMAR PUBLISHERS INC. ®

KALAMAZOO VALLEY
COMMUNITY COLLEGE
WITHDRAWN
LIBRARY

JAN 2 4 2000

NOTICE TO THE READER

Purchaser does not warrant or guarantee any of the products described herein or perform any independent analysis in connection with any of the product information contained herein. Publisher does not assume, and expressly disclaims, any obligation to obtain and include information other than that provided to it by the manufacturer.

The reader is expressly warned to consider and adopt all safety precautions that might be indicated by the activities described herein and to avoid all potential hazards. By following the instructions contained herein, the reader willingly assumes all risks in connection with such instructions.

The publisher makes no representations or warranties of any kind, including but not limited to, the warranties of fitness for particular purpose or merchantability, nor are any such representations implied with respect to the material set forth herein, and the publisher takes no responsibility with respect to such material. The publisher shall not be liable for any special, consequential or exemplary damages resulting, in whole or in part, from the readers' use of, or reliance upon, this material.

Cover Credits:
Cincinnati Milacron (Note: Safety equipment may have been
removed or opened to clearly illustrate products and must be
in place prior to operation.)
Anilam Electronics Corporation
General Electric Company (GE FANUC Automation)
Cover design by Juanita Brown

Delmar Staff:
Executive Editor: Wes Coulter
Developmental Editor: Lisa A. Reale
Project Editor: Eleanor Isenhart

Design Coordinator: Susan C. Mathews
Production Coordinator: Sandy Woods

For information address Delmar Publishers Inc.
2 Computer Drive West, Box 15-015,
Albany, New York 12212-5015

Copyright © 1990 Delmar Publishers Inc.

All rights reserved. No part of this work covered by the copyright
hereon may be reproduced or used in any form or by any means—
graphic, electronic, or mechanical, including photocopying,
recording, taping, or information storage and retrieval systems—
without written permission of the publisher.

Printed in the United States of America
Published simultaneously in Canada by Nelson Canada
a division of The Thomson Corporation

10 9 8 7 6 5 4 3 2 1

Library of Congress Cataloging in Publication Data

Goetsch, David L.
  Advanced manufacturing technology / David L. Goetsch.
    p.   cm.
  ISBN 0-8273-3786-8   ISBN 0-8273-3787-6 (Instructor's guide)
    1. Manufacturing processes—Automation.   2. Computer integrated
  manufacturing systems.   I. Title.
TS183.G64   1990   89-17147
670—dc20   CIP

WITHDRAWN

CONTENTS

PREFACE

Advanced Manufacturing Technology was written in response to the need for a book that examines the new manufacturing technology that has resulted from the concepts of **automation** and **integration** in manufacturing. This book was designed to be used by students who have completed a study of or are otherwise familiar with basic manufacturing concepts, including machining, casting, forming, assembly, and joining and fastening. Each concept presented in **Advanced Manufacturing Technology** relates to one or more of these basic manufacturing processes and shows how the process has been affected by automation, integration, or other developments that have resulted from the rapid emergence of computer technology. The basic manufacturing processes remain the same. However, the way these processes are controlled and carried out has changed as a result of automation and integration. **Advanced Manufacturing Technology** shows how and where these changes have occurred.

WHO IS THIS BOOK FOR?

Advanced Manufacturing Technology can be used as a textbook in a variety of different educational settings. It was designed for use in community college, university, and technical school classrooms as well as in on-the-job and industrial training settings. Examples of students who will benefit from a study of this book include

- Manufacturing technology students
- Industrial technology students
- Engineering technology students
- Mechanical engineering students
- Industrial engineering students
- Drafting and design students

- Machinist students
- Technology education students
- Industrial arts education students
- Vocational teacher education students

Practitioners in manufacturing, design, and drafting who need technical updating in the various advanced manufacturing technology areas covered in this book will also find **Advanced Manufacturing Technology** useful.

SPECIAL FEATURES OF THE BOOK

Advanced Manufacturing Technology has a number of special features that enhance its usefulness as a learning aid. The most important of these are the following:

1. Complex technical material is presented in a simple, easy-to-understand format and in easily readable language.
2. Each chapter is a complete learning package that includes comprehensive coverage of the subject matter accompanied by up-to-date illustrations and photographs, a listing of key words and phrases, a comprehensive list of questions for further study and review, and a set of **real-world** case studies relating to the subject matter provided by the **Society of Manufacturing Engineers (SME)** from its **Manufacturing Insights**® series of videotapes.
3. A variety of other support materials has been provided by SME, including text materials and illustrations.
4. A comprehensive glossary of advanced manufacturing technology terms and phrases is included.
5. There is a comprehensive index for easy access to specific material.

THE SME CONNECTION

The Society of Manufacturing Engineers or SME is a nonprofit technical society dedicated to the advancement of scientific knowledge in the field of manufacturing and to applying its resources to research, writing, publishing, and disseminating information. Founded in 1932, SME has its world headquarters in Dearborn, Michigan.

Education is one of the key activities of SME. The society produces curriculum materials, textbooks, training aids, and videotapes. As a member of the Accreditation Board for Engineering and Technology

(ABET), SME has responsibilities for the accreditation criteria in Manufacturing Engineering, Manufacturing Engineering Technology, and related degree areas.

SME's Manufacturing Engineering Education Foundation, through partnerships with industry, stimulates development of new and existing manufacturing engineering and technology through funding for capital equipment, research, awards, and student, faculty, and curriculum development. Further information about SME is available from the society's headquarters:

> Society of Manufacturing Engineers
> One SME Drive
> P.O. Box 930
> Dearborn, Michigan 48121-0930

SME has provided numerous materials used in the development of this book, including case studies from its **Manufacturing Insights**® series of videotapes, illustrations of manufacturing equipment and systems, and text material as noted within the book.

MANUFACTURING INSIGHTS® *VIDEOTAPE SERIES*

The SME produces a series of videotapes on advanced manufacturing technologies and concepts under the collective title **Manufacturing Insights**®. The series is updated and added to continually as new technologies emerge and existing technologies develop further.

The subject matter for each videotape is different, but the format is the same. Each tape contains the following:

- **Introduction** to the featured technology's operating principles, benefits, and typical equipment requirements
- **Interviews** with industry experts covering everything from methods to cost justification
- **Case Studies** examining companies of all sizes and the way they use the technology to meet specific manufacturing needs
- **Summary** of expected trends and advances to help you prepare for the future

Companies featured in this series of videotapes include Boeing Military Airplanes, Lockheed Aeronautical Systems, General Motors Corporation, McDonnell-Douglas, Pixley-Richards, Hydro Line Mfg. Co., Buick Hydra-Matic, Stihl, Inc., Defiance Metal Products, 3M Corporation, RCA Picture Tube Division, K2 Skis, Toro Outdoor Products, Oster Division of Sunbeam Corporation, Pillsbury Company, Century Plastics,

General Electric Company, Kearfott Division of Singer, Delta Faucet, Northern Telecom, Vought Aerospace, Intel Corporation, IBM Corporation, the U.S. Postal Service, Harley-Davidson, and Ford Motor Company.

A partial list of videotapes available in the *Manufacturing Insights*® series with descriptions of each tape is included in the Appendix. Updated lists may be obtained from SME at the previously shown address.

ABOUT THE AUTHOR

Dr. David L. Goetsch is Dean of Technical Education at Okaloosa-Walton Community College (OWCC) and Director of the College's Center for Manufacturing Competitiveness (CMC), which is a technology transfer center for advanced manufacturing technologies. Dr. Goetsch and other CMC personnel work with manufacturing firms to help them stay competitive through the adoption and optimum use of appropriate advanced manufacturing technologies. The services provided to manufacturing firms by the CMC include updating and retraining of manufacturing personnel, updating of management personnel and key decision-makers on new and emerging manufacturing technologies, selection and pilot testing of new technological systems and equipment, and setup and optimization of advanced manufacturing processes.

The CMC is the State of Florida's representative in the Southern Technology Council's Consortium for Manufacturing Competitiveness, a twelve-state consortium of community colleges, universities, and technical colleges that has technology transfer in the manufacturing arena as its main purpose.

ACKNOWLEDGEMENTS

The author wishes to thank the following people for their valuable assistance in reviewing this book. Through their input, the book has been made much better than it otherwise might have been. Robert E. King, Manager, Publications Development Department, Reference Publications Division, Society of Manufacturing Engineers; James F. Fales, Ed.D., CMfgE, Professor and Chairman, Department of Industrial Technology, Ohio University, Wallace Pelton, Texas State Technical Institute; and William M. Spurgeon, Director, Manufacturing Systems Engineering Program, The University of Michigan-Dearborn. The author also acknowledges the invaluable assistance of Faye Crawford in transcribing and word processing the manuscript.

Before beginning a study of the ten chapters that make up the main body of **Advanced Manufacturing Technology**, students can benefit from a discussion of the way the specific advanced manufacturing technologies included in the book were selected and how they fit together. The computer revolution and the related concepts of automation and integration have given rise to several new manufacturing technologies that affect the way basic manufacturing processes are controlled and performed. The advanced manufacturing technologies covered in this text are those that have the greatest effect on the way products are produced in the modern manufacturing plant.

Chapter 1 shows the way the computer and other related technological developments such as the robot have changed the traditionally most labor-intensive component in manufacturing, the assembly process. Through the use of computers, robots, parts feeders, conveyor systems, and other advanced technologies, the assembly process is evolving from a labor-intensive process into a technology-intensive one. Chapter 1 explains the way assembly processes are automated, where automated assembly is appropriate, and where it is less appropriate. It also explains the various types of systems and equipment used in automated assembly settings.

Chapter 2 explains how automated guided vehicles (AGVs) have changed the way manufacturing materials are handled and transported in some manufacturing settings. Originally developed to handle materials in settings that could be hazardous to human workers, AGVs are now being used in a variety of materials handling situations. The AGVs are changing the way materials are loaded, transported, and unloaded in manufacturing plants. This chapter explains how the technology developed, how it has evolved over the

Introduction

Automated Assembly

Automated Guided Vehicles

CAD/CAM

Numerical Control

Industrial Robots

Lasers in Manufacturing

Programmable Logic Controllers

Flexible Manufacturing

Computer Integrated Manufacturing

Other Related Technologies

years, how it is being used in modern manufacturing plants, how AGVs are guided and controlled, how AGV systems are managed, and appropriate, as well as less appropriate, applications of AGVs.

Chapter 3 covers the most widely recognized advanced manufacturing technology, computer-aided design/computer-aided manufacturing (CAD/CAM). This chapter shows how the design process has been automated in such a way as to eliminate the "wall" that has traditionally separated the design and production components of a manufacturing plant. This chapter shows how, through the sharing of a common database, the design and production components of a manufacturing plant can work together more closely and enjoy better two-way communication. Hardware, software, and related processes are covered in this chapter.

Chapter 4 covers one of the most widely applied advanced manufacturing concepts, numerical control. The concept is covered from its beginnings when paper tape was the programming medium to today's modern distributed numerical control systems. This chapter covers numerical control as a general concept, computer numerical control, direct numerical control, and distributed numerical control. Special attention is given to the emergence of the personal computer and the effect it has had on numerical control of manufacturing processes and systems.

Chapter 5 covers another widely recognized advanced manufacturing technology, the industrial robot. Industrial robots are being used in materials handling, assembly, joining and fastening, surface coding, and a variety of other settings in modern manufacturing plants. This chapter covers the different types of robots and their most appropriate applications. It also covers the operation and control of industrial robots. Other topics include sensors, machine vision as it relates to robots, and end-effectors.

Chapter 6 covers the many uses of the laser in modern manufacturing plants. For many years, the laser was considered a solution looking for a problem. In manufacturing, the laser found its problem and it has provided a number of solutions. This chapter covers the laser as a general concept as well as its several applications in modern manufacturing.

Chapter 7 explains the similarities between personal computers and modern programmable logic controllers. Logic controllers were originally hardwired boards that had to be set up for a specific process and then torn down and reconfigured each time the process changed. The emergence of programmable integrated circuitry allowed logic controllers to evolve from hardwired devices to programmable devices.

Chapter 8 shows how the advanced manufacturing technologies and concepts covered in this book can be tied together in integrated systems and sales. Flexible manufacturing is explained as a general concept. Flexible manufacturing systems and manufacturing cells are also explained, as are the similarities and differences between the two.

Chapter 9 covers the concept that is the integration and culmination of all of the various advanced manufacturing technologies and concepts. Computer-integrated manufacturing (CIM) is not just an advanced manufacturing concept, it is a whole new way of doing business in the manufacturing arena. With CIM, all of the various components and processes involved in designing, producing, marketing, delivering, and maintaining products are tied together or integrated via computers. This means that in a true CIM setting, there will be a shared database among all components of a manufacturing firm and each component will have direct access to that database. This chapter revolves around the "CIM wheel" developed by the Society of Manufacturing Engineers. Through it, students will see how all of the various manufacturing processes and related processes in manufacturing can be tied together into one comprehensive integrated system.

Chapter 10 covers several related advanced manufacturing concepts including the personal computer in manufacturing, statistical process control, computer-aided process planning, and computer-aided quality control.

In addition to the related concepts covered in Chapter 10, within the other nine chapters, a variety of other related advanced manufacturing concepts are covered. Included is coverage of such concepts as machine vision, sensors, artificial intelligence, manufacturing automation protocol, just-in-time delivery, initial graphics exchange specification, and technical and office protocol. These concepts are presented within the chapters where they apply rather than in separate chapters so that students can see that they relate to the various technologies covered in this book and are not stand-alone, independent concepts.

At the end of each chapter, case studies provided by the Society of Manufacturing Engineers from the *Manufacturing Insights®* series of videotapes are included. These case studies cover "real-world" situations where actual manufacturing companies have adopted the advanced manufacturing technology covered in the chapter. Representatives from these manufacturing firms tell in their own words of the effect the subject advanced manufacturing technology has had on their company. Students are encouraged to read the case studies carefully and relate them back to the material covered in the body of the chapter. An ideal approach is to study each chapter in this book and to watch the related videotape from the *Manufacturing Insights®* series.

A complete list of videotapes in the *Manufacturing Insights®* series produced by the Society of Manufacturing Engineers and a description of the contents of each tape is included in the Appendix. The various tapes contained in this series are excellent supplements to this text.

Major Topics Covered

- Automated Assembly Defined
- Current State of Automated Assembly
- Rationale for Automated Assembly
- Criteria for Automating Assembly Processes
- Inhibitors of Automated Assembly
- Advantages of Automated Assembly
- Impact of Design on Automated Assembly
- Control of Automated Assembly Systems
- Assembly Systems
- Principal Components of Robot Assembly Systems
- Parts Feeders in Automated Assembly Systems
- Transfer Lines
- Joining and Fastening in Automated Assembly Systems
- AGVs in Automated Assembly Systems
- Machine Vision in Automated Assembly
- Artificial Intelligence and Automated Assembly
- *Case Study: Remmele Engineering*
- *Case Study: Vibromatic*

Chapter One

Automated Assembly

Automated Guided Vehicles

CAD/CAM

Numerical Control

Industrial Robots

Lasers in Manufacturing

Programmable Logic Controllers

Flexible Manufacturing

Computer Integrated Manufacturing

Other Related Technologies

AUTOMATED ASSEMBLY DEFINED

Assembly is the process by which parts and subassemblies of manufactured products are put together to form the finished product. The assembly process varies according to the type of product being assembled and the parts and subassemblies that it comprises. However, the types of tasks typically included in the assembly process are the following (Figure 1–1):

1. fastening
2. inspection
3. acceptance or rejection of parts
4. labeling
5. packaging (where appropriate)
6. final preparation

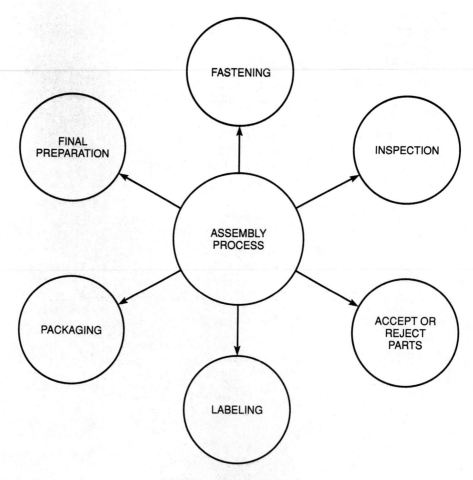

FIGURE 1–1 Assembly process.

As can be seen from Figure 1–1, the assembly process involves any and all tasks necessary to transform parts and subassemblies into a finished product. **Automated assembly**, then, can be defined as the process in which assembly of manufactured parts and subsystems is fully accomplished using computer-controlled machines rather than manual processes. The principal machines used in automated assembly are the computer and the robot. However, several other computer-controlled machines and systems such as automated guided vehicles (AGVs), parts feeders, and conveyor systems may also be used.

CURRENT STATE OF AUTOMATED ASSEMBLY

Automating the assembly process is complex and difficult, much more so than automating other processes such as machining, where only a small number of tasks are involved. For example, an automated machining center that drills, turns, and mills manufactured parts, in spite of its flexibility, still only performs the task of material removal. Assembly, on the other hand, could involve such radically different tasks as holding, fastening and joining, inspection, labeling, and packaging. For this reason, automated assembly has yet to achieve the wide-scale adoption that other automated processes now enjoy. Manual assembly still dominates in the workplace. Two things have held back the broad acceptance and wide-scale use of automated assembly: (1) insufficient flexibility of automated systems and (2) a lack of intelligence on the part of automated assembly systems.

The flexibility problem means that manufacturing firms using automated assembly processes have not been able to easily adapt when the market requires changes in the design of their product. The lack of machine intelligence means that automated assembly machines have not been able to make the many and varied types of decisions that a human assembler can make instantaneously. However, advances in technologies are beginning to overcome the problems of inflexibility and insufficient intelligence in automated assembly systems. Prominent among these technological developments is **vision** technology. This rapidly developing technology is allowing industrial robots to come much closer to imitating the intelligence of the human worker. As this and other related technologies continue to develop, so will the wide-scale adoption of automated assembly.

RATIONALE FOR AUTOMATED ASSEMBLY

As difficult as it is to automate the assembly process, one might question why manufacturing firms would attempt to do so in the first place. Such questions are usually asked when manual processes are being automated. It is a good idea to ask such questions because automation as a concept is not inherently better than any other concept. It is also not the right answer in every case. However, there are several good reasons behind the interest of manufacturing firms in automating the assembly process (Figure 1–2):

1. need to reduce high labor costs of manual assembly processes
2. need for better uniformity
3. need for greater capabilities
4. need to reduce in-process inventory
5. need for higher productivity
6. need for faster response to orders
7. need for higher reliability of product and process
8. need for consistent levels of quality

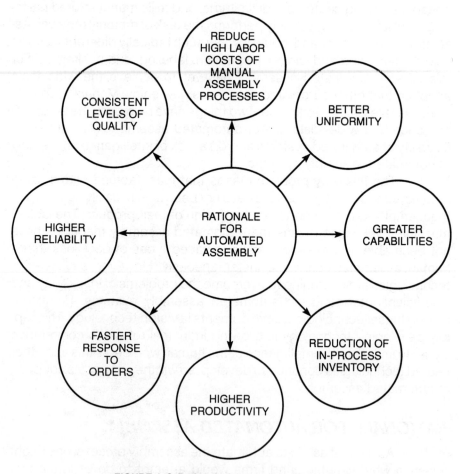

FIGURE 1–2 Rationale for automated assembly.

Manual assembly is a labor-intensive process. As a result, it drives up the cost of the finished product due to high direct labor costs. High direct labor costs, in turn, make it difficult for a manufacturer to remain competitive in an increasingly competitive international marketplace. By automating the assembly process, manufacturers can bring down the direct costs associated with assembly.

Another reason for automated assembly is uniformity. Consumers of manufactured products look for uniformity of the product. They want to know that they can expect the same product each time it is ordered. This is the same rationale used for automating other manufacturing processes, such as machining. Computer-controlled, automated machines are able to produce a greater degree of uniformity than their manual counterparts.

The next reason is enhanced capabilities. The machines can be used to inspect, fasten, accept or reject parts, and apply labels as part of an automated assembly system, are more accurate and dependable than human assemblers, and produce a more uniform result. The greater capabilities of automated assembly systems lead to the next reason behind the interest in automated assembly systems; reduction of in-process inventory. Greater capabilities in such critical areas as accuracy, dependability, and speed allow manufacturers to reduce the amount of in-process inventory, thereby reducing their fixed overhead. This, in turn, reduces the eventual costs of the manufactured product and makes the manufacturer more competitive.

Higher productivity is a primary reason behind the automation of any manufacturing process, including the assembly process. The higher productivity of some computer-controlled, automated machines compared with human operators is well established. This higher productivity, coupled with the greater capabilities of automated assembly systems, leads to the next reason for the interest in automated assembly; faster response to orders. Automated assembly systems allow manufacturers to respond faster to customer orders. Another reason is the higher reliability of automated assembly systems when compared with manual systems, which is well documented. This is a reason that applies to automation of any manufacturing process.

A final reason for the interest in automated assembly processes is in their ability to provide consistent levels of quality in finished products. Inconsistency is an inherent human weakness and is consistency an inherent machine strength. Consistency in the quality of the finished product is an absolute necessity in a competitive international market place.

Since automated manufacturing systems are still in their infancy, this rationale may not apply to systems currently on the market.

Rather, they apply to the systems that will emerge as advances in technology continue to move us toward automated assembly systems that fully exploit their theoretical potential.

CRITERIA FOR AUTOMATING ASSEMBLY PROCESSES

No manufacturing firm should go blindly down the path of automation. Automated assembly, like all automated manufacturing processes, is appropriate in some settings and not in others. Therefore, there must be some criteria for making a decision as to whether or not to automate the assembly process. Before seeking to automate the assembly process, manufacturers should understand where automated assembly is appropriate and where it is not. The most appropriate application of automated assembly is in producing products that are small and stable in their design. Even the simplest design change can require the system to be reprogrammed; retrofitted with new fastening, inspection, labeling, or packaging equipment; or completely overhauled.

Small products do better in automated assembly systems for very practical reasons. The principal component in an automated assembly system is often a robot. Although this technology is developing rapidly, electrical robots are still not able to lift and manipulate extremely heavy parts or subassemblies.

Figure 1–3 is a questionnaire that can be used to determine the feasibility of automating the assembly process in a given manufacturing setting. Note that items 1 through 9 on the checklist deal with the design of the product and the existing assembly process. Items 10 and 11 deal with the availability of the types of personnel needed to operate and maintain automated assembly systems. These are both critical criteria because even the best automated assembly system is like a race car without a driver and crew if there are no competent, qualified technical personnel available to operate and maintain it. Items 12, 13, and 14 are management and labor criteria. Any manufacturing firm that can respond "yes" to all of the criteria set forth in this checklist can automate their assembly process with a reasonable expectation of success. Companies that cannot respond "yes" to these fourteen criteria might be better off trying to improve their manual assembly process.

INHIBITORS OF AUTOMATED ASSEMBLY

Automated assembly is still an emerging technology that will undergo much additional development before it is a widely accepted

These questions will help manufacturing personnel decide whether a transition from manual to automated assembly is appropriate in a given setting.

Criteria	Response	
	Yes	No
1. Is the assembly simple and made of few parts?	_____	_____
2. Do the parts lend themselves to automated inspection?	_____	_____
3. Are the parts light and easy to handle?	_____	_____
4. Is the product relatively free from design changes?	_____	_____
5. Is the product free from complex joining and fastening procedures?	_____	_____
6. Are manual assembly times long?	_____	_____
7. Are manual assembly costs high?	_____	_____
8. Does the manual process have a high rejection rate?	_____	_____
9. Is the product needed in large volume?	_____	_____
10. Are qualified personnel available to operate an automated assembly system?	_____	_____
11. Are qualified personnel available to maintain an automated assembly system?	_____	_____
12. Will a conversion lead to labor problems?	_____	_____
13. Is top-level management knowledgeable enough of automated assembly to have a responsible decision?	_____	_____
14. Is top-level management committed to automated assembly enough to make the front-end investment?	_____	_____

FIGURE 1–3 Decision-making checklist.

concept. Before automated assembly can become the norm in manufacturing, three major inhibitors will have to be overcome (Figure 1–4):

1. inability to handle variations in parts
2. expensive initial investment for hardware and software
3. inability to self-correct insertion problems

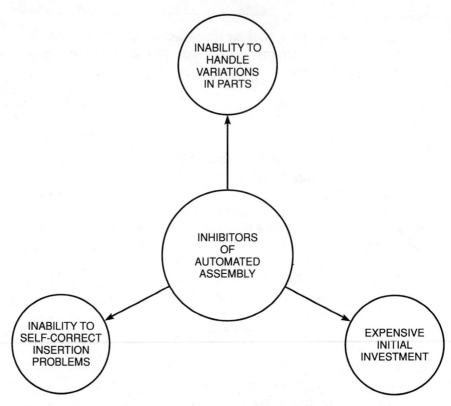

FIGURE 1–4 Inhibitors of automated assembly.

Even parts that are designed to be identical can vary slightly in their characteristics. At this point in the development of automated assembly systems, such variations cannot be accommodated. Parts are designed to be manufactured within certain specified tolerance zones for size, shape, and the location of geometric features within the part. Even the tightest tolerances specified for the best, most modern machine tools allow minor variations. However, even the smallest variations can cause an automated assembly system based on current technology to improperly reject a part, fail in an attempt to insert another part, or fail to properly fasten parts together.

The high initial investment required to purchase an automated assembly system is a major inhibitor, as it has been with all automated manufacturing technologies. As is always the case, when a new technology is emerging, hardware and software costs can be prohibitive. However, as the technology continues to develop and competition among vendors brings costs down, this inhibitor will be overcome.

The final major inhibitor of automated assembly is the inability to self-correct insertion or fastening problems. When a human operator places a nut on a bolt or inserts a male part into a female part incorrectly, the operator is able to adjust immediately, thereby correcting the error. At this point in the development of automated assembly technology, self-correction is still a problem. Advances in machine vision and robot end-effector technology will eventually solve the problem. However, until such advances are realized, self-correction will remain a major inhibitor of the wide-scale adoption of automated assembly.

ADVANTAGES OF AUTOMATED ASSEMBLY

At this point in its development, automated assembly is experiencing all of the problems that other new automated manufacturing technologies have experienced. As has been the case with other technological developments in manufacturing, automated assembly technology will continue to develop and advance to the point that its full potential can be realized. When that happens, automated assembly systems will offer several advantages over manual assembly. Figure 1–5 is a checklist of advantages that will result from automated assembly once the concept is completely developed and realizes its full potential.

This is a list of *potential* advantages of automated assembly as compared with the manual approach. These advantages are qualified as being potential because they are based on the full development of automated assembly technology rather than the current state of the art.

1. Better product quality
2. Higher levels of consistency and uniformity
3. Fewer rejected parts
4. Higher level of production reliability
5. Reduced warranty costs due to higher levels of quality, consistency and uniformity
6. Improved productivity and resultant decreased manufacturing costs
7. Improved safety in dangerous assembly environment
8. Reduced inventory costs
9. More efficient production scheduling
10. Reduced floor space needs

FIGURE 1–5 Advantages of automated assembly.

IMPACT OF DESIGN ON AUTOMATED ASSEMBLY

The success of automated assembly depends in large measure on close cooperation and collaboration between design and manufacturing personnel. The design of a part is the key to whether that part will lend itself to automated assembly. Parts that do not lend themselves to automated assembly must either be assembled manually or redesigned for automated assembly. Traditionally, parts have been designed for this functionability of the part. It may now be necessary for design engineers to design for both part function and ease of automated assembly.

When attempting to design a part for ease of automated assembly, design engineers should consider design features that will make the part easy to handle, simple to orient, simple to fasten, machinable to realistic tolerances, and have sufficient clearances to allow for the full movement of assembly tools. Design engineers can enhance the assembly process even further by completely eliminating fastening as a part of the process whenever possible.

This marriage between design engineers and manufacturing engineers is not limited only to assembly. It is part of a larger concept that has developed out of necessity in order for manufacturers to maintain and improve their position in a competitive international market. Engineers who designed products that would be manufactured and assembled manually did not have to be concerned with the processes used to manufacture and assemble the finished product. Those processes would be designed around the part or parts in question. However, with the advent of automated manufacturing processes, this is no longer feasible. Manufacturing and engineering personnel have found that to gain the full advantages of automated manufacturing technologies, it is necessary to design for both function and automation. This concept is sometimes referred to as **automation-friendly design**.

Henry W. Stoll, in a paper written for the Society for Manufacturing Engineers entitled "Automation: Teach and Test for Good Product Design," described a three-step process for automation-friendly design. The three steps are the following:

1. Identify those aspects of the process for making the product that will be easy to automate.
2. In component design, make all decisions with automated assembly in mind.
3. Integrate the design of automated systems with the design of productivity improvement strategies to achieve the optimum balance therein.

The key in helping design engineers develop automation-friendly designs lies in compiling a list of "designing for automation" principles that can be applied by engineers. This list of principles, which should be developed in conjunction with manufacturing engineers, is then used by design engineers as a guide in developing automation friendly designs.

CONTROL OF AUTOMATED ASSEMBLY SYSTEMS

Automated assembly systems may be controlled by programmable logic controllers, robot controls, computers, and/or communication systems (Figure 1–6). The type of control mechanisms used with automated assembly systems depends on the level of sophistication of the system. There is a **control hierarchy** associated with automated assembly systems. The higher up one goes from the base of the hierarchy, the more sophisticated the control becomes. The control hierarchy for automated assembly systems is shown in Figure 1–7. The first and least sophisticated level is made up of the individual assembly machines, which are single machines that perform a specific, highly specialized assembly

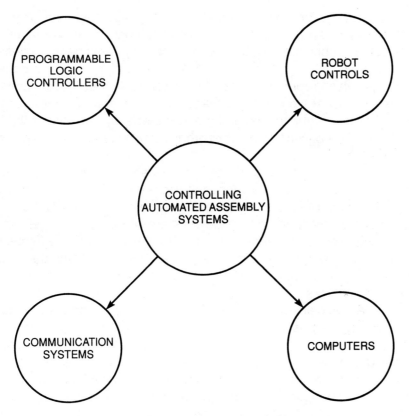

FIGURE 1–6 Controlling automated assembly systems.

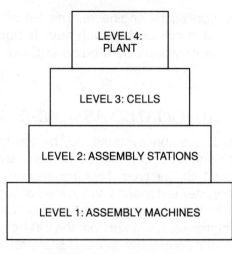

FIGURE 1-7 Control hierarchy.

task. If one were to stop at this level, only a minimum of control would be required.

Level 2 groups individual assembly machines into assembly stations, which are two or more machines that perform two or more specialized assembly tasks. At this point, control systems begin to become more extensive and sophisticated. Assembly stations can be combined into cells, which represent level 3 in the control hierarchy. An assembly cell is a group of workstations that perform several difficult assembly tasks. The top level in the hierarchy is the plant level. The plant is the total of all the assembly cells and other production systems used to produce the product.

At the machine level, control devices are interfaced with such mechanical apparatus as switches, buttons, solenoids, motor starters, and a variety of other sensing and driving devices. At the station level, more sophisticated controlling devices are required. Intelligent input-output modules for connecting the control device to the machines that make up the assembly stations is necessary. More sophisticated motion control systems are needed, as are programmable controllers capable of logic-solving tasks.

At the cell level, control shifts from **real-time operations** to supervisory operations. Highly sophisticated computers communicate directly with programmable logic controllers or computers at the station level to direct their operations. This concept is known as distributed control. Computers or other control devices at the cell level are used as host computers to provide data to control devices at the assembly level. Programs stored at the cell level can be downloaded to computers or programmable controllers at the station level. At that point, the station-level machines can operate without further interaction with controllers at the cell level. Once the required operations have been performed at the station level, the programs can be returned to the host computer at the cell level.

Control at the highest level, the plant level, covers more than the assembly process. Control at this level integrates all of the other manufacturing operations that precede and follow the assembly process. This level of control is the most sophisticated and the most difficult to accomplish.

An issue that is not yet resolved in the control of automated systems and processes in a manufacturing plant is communication among and between all of the various islands of automation. An island of automation is a special group of automated machines that work together but have no direct communication with other machines and systems outside their group, hence the term **island of automation**. A manufacturing plant may have automated its design process, machining processes, and assembly processes. However, the hardware and software in each of these islands of automation may have been produced by different vendors. A lack of standardization in the areas of hardware and software makes control at the top of the control hierarchy difficult to accomplish.

This problem is being solved by an effort known as the **manufacturing automation protocol (MAP)**. A MAP program is an approach to standardization between and among vendors that is based on the development of communication standards coupled with the use of very large scale integration (VLSI) technology. The development of the MAP standardization program is a major step in accomplishing the complete control and integration of islands of automation within a manufacturing plant.

ASSEMBLY SYSTEMS

There are numerous different types of machines and systems available for use in automated assembly operations. Each falls into one of the following categories:

1. single-station
2. synchronous
3. nonsynchronous
4. flexible

Regardless of the type of system used, there is a set of basic equipment that is common to all. These common components include the following:

1. workholding devices
2. transfer or indexing mechanisms

3. parts feeding devices
4. orienting devices

In addition to these basic equipment requirements, automated assembly systems must have a means for removing defective parts, controlling noise, ensuring safety, and protecting the environment.

Single-Station Systems

In assembly operations where a high volume of a few parts are handled, single-station systems are widely used. An example of such a setting would be when many parts are assembled into one base unit. Single-station machines are occasionally grouped to form multiple-station assembly systems.

Synchronous Assembly Systems

Synchronous or indexing assembly systems come in three types (Figure 1–8):

1. dial or rotary
2. in-line
3. carousel

Synchronous systems are characterized by all parts or pallets carrying parts undergoing synchronous movement: They all move at the same time and for the same distance. This means the intervals between operations (indexing intervals) must be based on the slowest operation. A disadvantage of synchronous systems is that a breakdown anywhere along the line shuts down the whole line.

Synchronous systems are used primarily in settings involving small, lightweight parts. They are capable of high-speed, high-volume production. A rate of 4000 assemblies per hour is not uncommon with synchronous systems.

Nonsynchronous Assembly Systems

Nonsynchronous assembly systems are characterized by parts and pallets carrying parts moving at different times and for different distances. Nonsynchronous assembly systems come in three configurations (Figure 1–9):

1. in-line
2. carousel
3. other

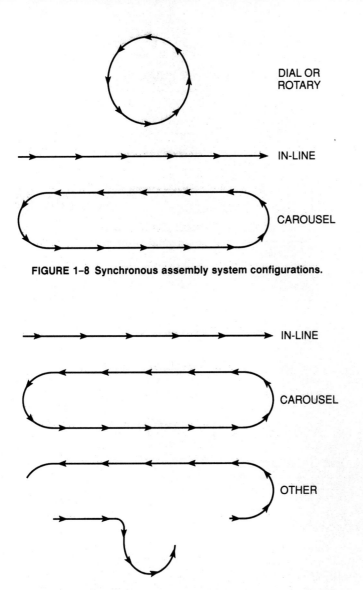

DIAL OR
ROTARY

IN-LINE

CAROUSEL

FIGURE 1–8 Synchronous assembly system configurations.

IN-LINE

CAROUSEL

OTHER

FIGURE 1–9 Nonsynchronous assembly system configurations.

Such systems are used in settings where the time required to complete operations at different stations varies widely. Nonsynchronous systems cannot match the production rates of synchronous systems, but they make up for this shortcoming through a higher degree of flexibility. They are flexible because individual stations can operate independently and parts can be supplied on demand. This allows for a combination of machine and human operation. Because of this, a breakdown at one point on the line does not shut down the entire line.

The typical assembly per hour rate for a nonsynchronous system is approximately 1400. Figure 1–10 shows a nonsynchronous system for assembly brake backing plates.

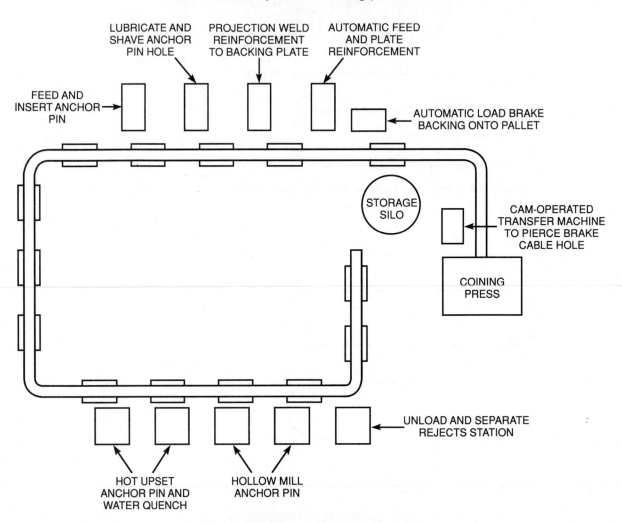

FIGURE 1–10 Nonsynchronous assembly system.

Flexible Assembly Systems

Automated assembly has the same need for greater flexibility as other automated manufacturing processes. Future demands will be less and less for high-volume long assembly runs and more and more for flexibility. A flexible assembly system is one that can handle a wide variety of products in smaller lot sizes. The advent of computer control, robots, and AGVs has made flexible automated assembly a reality.

PRINCIPAL COMPONENTS OF ROBOT ASSEMBLY SYSTEMS

Automated assembly systems can vary widely in configuration, depending on the needs of the individual manufacturing setting. However, regardless of the setting, the three principal components of a robot automated assembly system are robots, controllers, and parts feeders (Figure 1–11).

Robots used in automated assembly systems fall into one of three categories, depending on the type of power used to operate them:

1. electrically powered
2. pneumatically powered
3. hydraulically powered

Pneumatically and hydraulically powered robots can handle heavier parts than electrically powered robots. However, electrically powered robots are more accurate and better able to deal with small parts. Figures 1–12 and 1–13 are examples of modern industrial robots that might be used in an automated assembly system.

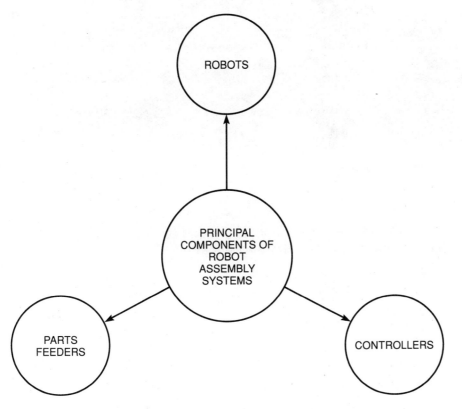

FIGURE 1–11 Components of a robot assembly system.

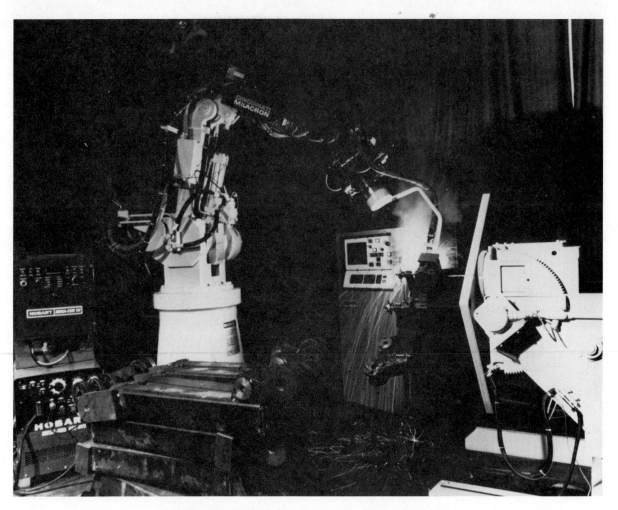

FIGURE 1–12 Assembly robot. *Courtesy of Cincinnati Milacron.*

Controllers used in automated assembly systems are of four basic types:

1. programmable logic controllers
2. motion controls
3. computers
4. communication systems

The type of controllers used in automated assembly systems is dictated by the level of control in the control hierarchy, as explained in the previous section.

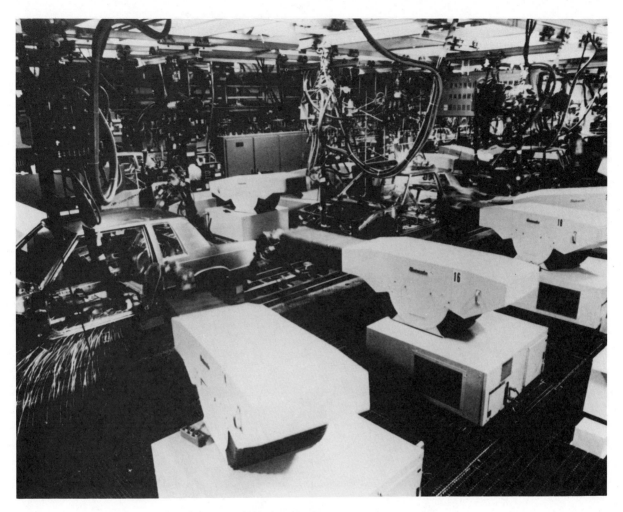

FIGURE 1-13 **Assembly robot.** *Courtesy of Cincinnati Milacron.*

PARTS FEEDERS IN AUTOMATED ASSEMBLY SYSTEMS

Parts feeders get less attention than robots and controllers, but they can be an important part of an automated assembly system. Figure 1-14 illustrates some of the typical parts feeder nomenclature. There are several types of parts feeders available including rotary bowl, orbital bowl, vibrator bowl, straight-line vibrator, and belt conveyor orienters, (Figure 1-15). A bowl-type feeder is shown in Figure 1-16 and

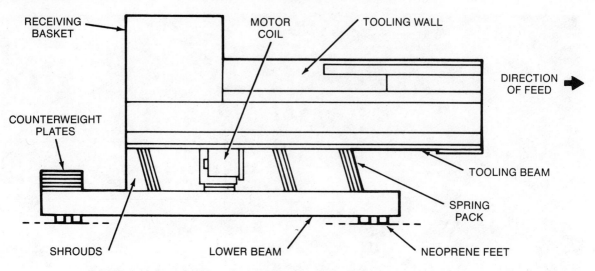

FIGURE 1-14 Parts feeder nomenclature. *Courtesy of Spectrum Automation Company.*

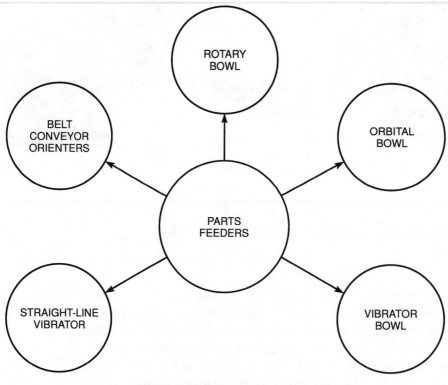

FIGURE 1-15 Types of parts feeders.

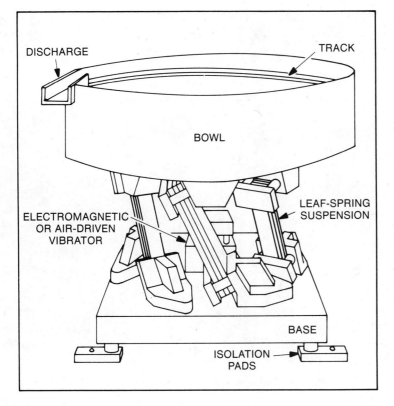

DISCHARGE

TRACK

BOWL

ELECTROMAGNETIC
OR AIR-DRIVEN
VIBRATOR

LEAF-SPRING
SUSPENSION

BASE

ISOLATION
PADS

FIGURE 1–16 Vibrator bowl feeder. *Courtesy of Society of Manufacturing Engineers.*

a belt conveyor type feeder is shown in Figure 1–17. Of the various types of parts feeders used in automated assembly systems, the rotary bowl feeder is the simplest. Its use is normally limited to settings involving a low to moderate production rate and parts that are simple in design. Straight-line, vibratory, and belt conveyor orienter parts feeders are more sophisticated.

In addition to feeding parts, parts feeders are capable of tasks such as multiple path distribution, live feeding, quick changing, and bolted-on tooling plates to accommodate parts families. Figures 1–18, 1–19, and 1–20 are examples of widely used parts feeders.

TRANSFER LINES

A **transfer line** is a specialized manufacturing system that can be used for a variety of purposes, one of which is automated assembly. A transfer line typically has four major components: (1) machine tools, (2)

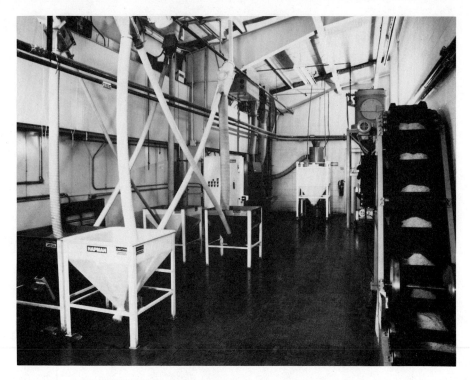

FIGURE 1–17 Conveyor system. *Courtesy of Hapman Conveyors.*

FIGURE 1–18 Parts feeder. *Courtesy of Spectrum Automation Company.*

FIGURE 1–19 Parts feeder. *Courtesy of Spectrum Automation Company.*

FIGURE 1–20 Parts feeder. *Courtesy of Spectrum Automation Company.*

assembly machines, (3) a conveyor system, and (4) loading and unloading equipment. Secondary equipment includes parts feeders, orientation devices, and buffer storage devices. In-process gauging and inspection devices may also be included.

Figure 1–21 is an example of a transfer line that could be used in an automated assembly setting. Such a transfer line would see parts fed onto the system, oriented, machined, inspected, assembled, and unloaded in one continuous operation controlled by a central supervisory computer.

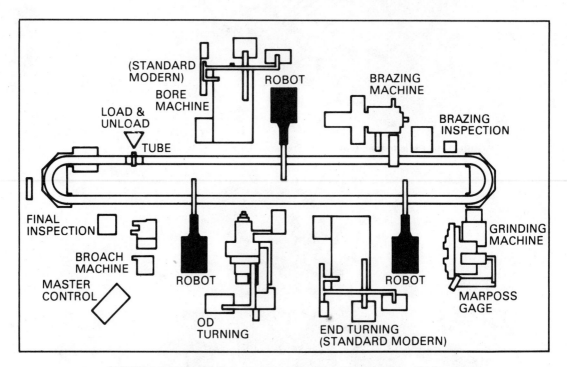

FIGURE 1–21 Transfer line. *Courtesy of Society of Manufacturing Engineers.*

JOINING AND FASTENING IN AUTOMATED ASSEMBLY SYSTEMS

Joining and **fastening** methods can be divided into four categories:

1. mechanical fastening
2. welding
3. brazing and soldering
4. adhesive joining

Most of the various joining and fastening methods in these four broad categories can be accomplished through automated assembly processes. However, note that joining and fastening are the most difficult processes to accomplish on automated systems.

Mechanical fastening includes threaded fasteners, riveting, eyeleting, stitching, stapling, and the use of snap rings, pins, and washers. Welding methods include arc, resistance, electron-beam, laser-beam, and ultrasonic welding. Soldering and brazing stand alone as joining processes. Adhesive joining is accomplished using natural inorganic or synthetic adhesive materials.

A good rule of thumb to follow in automated assembly systems is to eliminate joining and fastening operations whenever possible. Joining and fastening are the two most difficult processes for automated assembly technology to handle. The feasibility of automated assembly increases in those settings that do not require joining and fastening. However, joining and fastening are so indigenous to assembly that they cannot always be eliminated from the process.

By its very nature, the assembly process usually requires some joining and fastening. If joining and fastening cannot be eliminated from the assembly process, the amount of joining and fastening required should be decreased as much as possible. In addition, parts should be designed in such a way as to use the simplest, easiest to accomplish types of joining and fastening methods. It is also important that fastening methods be adopted that require no special tools. Figure 1–22 shows several methods of joining and fastening that work well in automated assembly settings. Figure 1–23 shows the types of joining and fastening methods that should be avoided in automated assembly systems.

You can see that the types of joining and fastening methods that should be avoided are the more cumbersome and difficult types to accomplish. It is difficult enough for a human operator to fasten clamps and screws and to place nuts on bolts with all the human flexibility available to them. Consider, then, the difficulty of accomplishing these tasks using robots.

Even when the joining and fastening methods used in an automated assembly setting can be limited to snap fits, interference fits, staking, spinning, or welding, it is best to accomplish joining and fastening at separate stations where inspection devices ensure the proper alignment of parts prior to the joining process taking place. Manufacturing firms that produce electromechanical products are finding that the laser can solve many of the joining and fastening problems they have traditionally experienced in welding and soldering operations.

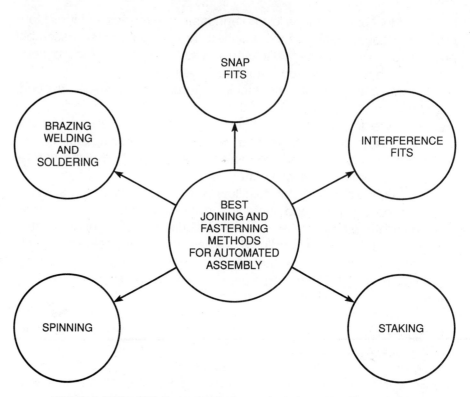

FIGURE 1–22 Best joining and fastening methods for automated assembly.

AGVs IN AUTOMATED ASSEMBLY SYSTEMS

Automated guided vehicles (AGVs) are rapidly becoming an important component of automated assembly systems. Early AGVs were referred to as wire-guided vehicles. They first began to appear in manufacturing settings in the early 1970s. They have been used more in other countries than in the United States since that time.

Rationale for AGVs

The AGVs are used primarily for materials handling. They are capable of carrying extremely heavy loads for long distances and were originally developed to reduce the costs associated with manufacturing by reducing the amount of labor required. Their first wide-scale use was in the manufacture of automobiles. Pioneered by Volvo, they were used as an alternative to the traditional assembly line. They allow for a more flexible approach to assembly that would also improve job satisfaction and worker motivation. Because of the flexibility and movement capabilities of AGVs, assembly jobs could be structured in such a way

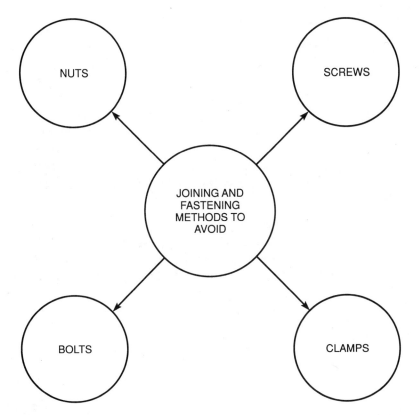

FIGURE 1-23 Joining and fastening methods to avoid in automated assembly.

that workers could communicate freely with one another, rotate jobs periodically, vary the rate of production where necessary, and have a greater influence over their working environment. The AGVs have done an excellent job in satisfying these purposes.

Now that the concept and technology are more highly developed and more sophisticated, AGVs can offer additional benefits in a manufacturing plant. With traditional assembly lines, setup and breakdown costs are high. Once an assembly line is set up, even minor changes are time consuming and expensive. However, with AGVs, changes can be made easily by revising a computer program and making variations to the AGV's guide path. The AGVs now are able to accommodate nonsequential assembly, which makes the assembly process even more flexible. Perhaps the most important rationale for AGVs has to do with the future of manufacturing. As manufacturers come to rely more and more on CAD/CAM for modeling and simulation of manufacturing processes, AGVs will improve in value. By their nature, AGVs lend themselves to computer modeling and simulation.

Advantages of AGVs in Automated Assembly

The AGVs offer several advantages over traditional materials and handling technologies (Figure 1–24). Because they are wire or tape guided and computer controlled, they are very flexible. Changes can be made at any time by simply revising a computer program and making variations in guide paths. Over the long term, AGV costs are low. Perhaps the most important advantages are that AGVs work well with robots and other automated manufacturing technologies and fit in well with the flexible manufacturing philosophy that will guide manufacturing processes in the future. They also offer several intangible cost benefits:

1. lower costs for line modification
2. lower costs due to inventory reduction and control
3. lower costs resulting from increased productivity
4. lower costs resulting from better quality control
5. lower costs resulting from more positive and more harmonious environment

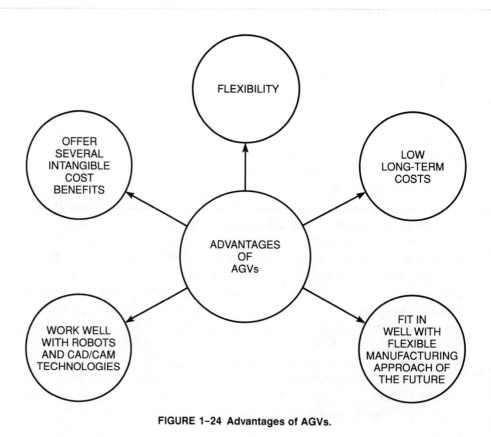

FIGURE 1-24 Advantages of AGVs.

MACHINE VISION IN AUTOMATED ASSEMBLY

Machine vision is perhaps the most important technological concept associated with automated assembly. The functional problem that has held back the full-scale development of automated assembly is the inability of key assembly components such as robots to make the many intelligent decisions necessary in an assembly setting. Machine vision is the process wherein a continuous flow of information is collected from visual sensing devices and fed back to machines to allow them to make intelligent decisions.

For example, a human assembly worker would see immediately if a part to be fastened was not properly oriented. The operator would then make a decision to properly orient and fasten the part. Without machine vision, however, automated assembly machines would not be able to observe that the part was not properly oriented and make the necessary intelligent decision to take appropriate action. Vision systems combine the technologies associated with computers and cameras to collect and process data that allows machines to undertake such tasks as inspection, sorting, and assembly.

Machine vision technology gives automated assembly machines several advantages over human operators using even the most advanced manual inspection techniques and tools:

1. the ability to inspect parts without making contact
2. better accuracy
3. better consistency
4. the ability to check 100% of the parts produced and assembled

Machine vision systems perform a variety of tasks in automated assembly systems. These tasks may be divided into three broad categories (Figure 1–25). Inspection tasks performed by machine vision systems in automated assembly systems include verification and flaw detection. Identification tasks include symbol and object recognition. Machine guidance tasks include object location and tracking.

ARTIFICIAL INTELLIGENCE AND AUTOMATED ASSEMBLY

An emerging technology that could revolutionize automated assembly is **artificial intelligence**. The most difficult problem to overcome in automated assembly is the inability of assembly systems to mimic such basic human capabilities as adjusting appropriately to differences in the size, shape, and orientations of objects. Artificial intelligence may be the answer to this dilemma.

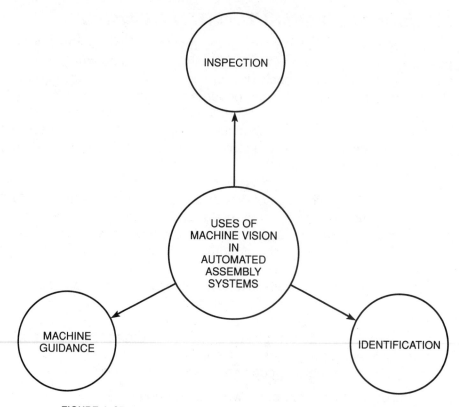

FIGURE 1-25 Uses of machine vision in automated assembly systems.

For example, an assembly worker whose job is to retrieve small parts from a feeder bin and insert them into the appropriate holes in a plate has many human attributes to assist him. These include sight, hand-eye orientation, reasoning abilities, logic, judgment, and experience.

If the assembly worker picks up a part that is not properly oriented or if the part does not properly seat on the first attempt at inserting, human capabilities allow the worker to adjust appropriately. Even the most sophisticated automated assembly systems cannot completely mimic these capabilities. Artificial intelligence is an attempt to increase the number of human characteristics computers and computer controlled systems can mimic.

Artificial Intelligence Defined

Artificial intelligence is the ability of a computer to imitate human intelligence and, thereby, make intelligent decisions. Computer-controlled systems that apply artificial intelligence to everyday settings are called **expert systems**.

In any discussion of artificial intelligence, there are several frequently used key words and phrases. Students of automated assembly should be familiar with the following terms:

Algorithm: A special computer program that will solve selected problems within a given time frame.

Early Vision: Computer calculations that allow systems to see by providing low-level data such as spatial and geometrical information.

Higher Level Vision: Computer calculations that allow systems to accomplish higher level tasks such as smart movement within an environment, object recognition, and reasoning about objects.

Knowledge Engineering: A process through which knowledge is collected from experts in a given field and converted into a computable format.

Neurocomputing: An approach to performing mathematical calculations on a computer that is based on the way the human nervous system operates.

Humans can make logical, reasoned adjustments in a work setting because they can quickly collect information, access it against the sum total of their human experience, and evaluate known relationships along various items of information. The science of artificial intelligence attempts to initiate this process with computers.

Humans attempt to create an experience base in computers by feeding them all the known information about a given subject. This information is then used by the computer in making decisions. This is why the concept called artificial intelligence. The computer does not really think; it simply searches its memory for the appropriate information. If the information is there, the computer uses it in making logical decisions. The key is in feeding the computer enough relevant information.

Historical Development of Artificial Intelligence

Figure 1–26 is a time line showing milestones in the development of artificial intelligence. In the 1950s scientists interested in artificial intelligence brought computers from number processors to symbol processors. This was a major step forward. The next step involved the development and use of algorithms for solving problems.

The 1960s saw the development of the **heuristic search**. This significantly reduced the amount of space through which a computer

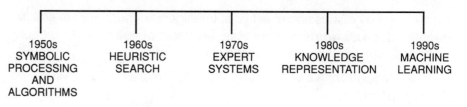

1950s	1960s	1970s	1980s	1990s
SYMBOLIC PROCESSING AND ALGORITHMS	HEURISTIC SEARCH	EXPERT SYSTEMS	KNOWLEDGE REPRESENTATION	MACHINE LEARNING

FIGURE 1–26 Artificial intelligence time line.

would search for solutions. Heuristics are rules of thumb that limit the size of the space searched by getting the computer in the ballpark from the outset. The one weakness of the heuristic search is that, although it locates a solution faster, the solution is not necessarily the best solution. However, a good heuristic search gives up only a little in finding a solution, while gaining a great deal in limiting the size of the search area.

The 1970s were characterized by the development of expert systems, which resulted from the realization of scientists that intelligence would not be achieved by searching through general space. They began to see the need for feeding task-specific information to the computer that would give it an experience base relating to a specific task domain. The result was the development of expert systems.

Expert systems have two key components: (1) an inference engine and (2) an experience or knowledge base. The first component controls the application of information contained in the second component. The development and use of expert systems raised questions about the way knowledge is represented.

Improving the ways knowledge is represented so that it is more explicit, yet more concise characterized the 1980s. The 1990s will see a rapid growth in the use of artificial intelligence in automated assembly and other manufacturing applications as the concept of machine learning evolves.

It is generally accepted that a key characteristic of human intelligence is the ability to learn. For artificial intelligence to reach its potential, the concept of **machine learning** must be fully developed. Machines learn in one of three ways:

1. parameter adjustment
2. concept formation
3. evolution of structure

Parameter adjustment is the most basic of the three approaches to machine learning. It involves adjusting the values of the parameters of a predetermined representation. It can affect only the values and not the structure of the predetermined representation. Concept formation involves grouping related objects into categories or groups. Evolution of structure makes use of neurocomputing. This involves the parallel activity of elements that are able to communicate the results of computations among themselves.

Artificial Intelligence in Automated Assembly

Artificial intelligence has applications in a number of manufacturing settings, one of which is automated assembly. In such a setting, artificial intelligence can be used to allow robots and other assembly systems to duplicate such human capabilities as vision and language processing. It can help improve the assembly skills of robots and other machines. Finally, artificial intelligence can improve the ability of information management systems used in automated assembly settings.

KEY TERMS

Assembly
Automated assembly
Vision
Automation-friendly design
Control hierarchy
Real-time operations
Island of automation
Manufacturing automation
     protocol (MAP)
Robot
Parts feeder
Transfer line
Joining

Fastening
Automated guided vehicle (AGV)
Machine vision
Artificial intelligence
Expert system
Algorithm
Early vision
Higher level vision
Knowledge engineering
Neurocomputing
Heuristic search
Machine learning

Chapter One REVIEW

1. Define assembly.
2. Define automated assembly.
3. List at least four categories of tasks that might be part of an automated assembly process.
4. What is vision technology?

5. List and explain four reasons for the interest in automating the assembly process.
6. What is the most appropriate application of automated assembly? Why it is the most appropriate application?
7. List and explain the three principal inhibitors of automated assembly.
8. How do design tolerances as to size, shape, and location of geometric characteristics affect automated assembly?
9. What is the difference between designing for functionability only and designing for assembly as well as functionability?
10. What is automation-friendly design?
11. List the three steps in developing an automation-friendly design.
12. Explain the four levels in the control hierarchy for automated assembly systems.
13. What is an island of automation?
14. Explain manufacturing automation protocol (MAP) and its potential impact on automated assembly.
15. What are the three principal components of an automated assembly system?
16. What are the three categories of robots used in automated assembly systems?
17. What are the four types of controllers used in automated assembly systems?
18. What is a parts feeder?
19. Explain why joining and fastening operations do not fit in well with automated assembly processes.
20. Explain the rationale for using AGVs in automated assembly.
21. Define the following terms:
    Artificial intelligence
    Expert system
    Early vision
    Neurocomputing
    Knowledge engineering

These case studies were provided by the Society of Manufacturing Engineers (SME). They are excerpted from the SME's *Manufacturing Insights*® series of videotapes. These case studies give students actual examples of the way the advanced manufacturing technologies covered in this chapter are being applied in "real-world" manufacturing settings.

As you read each case study, relate the examples cited to the material presented in the text of the chapter. This combination of textbook information and real-world examples will be particularly valuable to your understanding of advanced manufacturing technologies.

REMMELE ENGINEERING

Remmele Engineering Incorporated specializes in designing machines for unusual or difficult applications. They make automotive blade fuses for one of the leading suppliers in the industry. And they make them faster, cheaper, and of higher quality than formerly possible.

Assembling the fuses is a four-part process. Fusible link material is attached to the fuse body with resistance spot welds. A plastic window is precisely placed and welded ultrasonically. The fuses are cut and shaped, then tested and stamped for the proper amperage rating.

The process begins with fusable link material being fed from a 15-in. reel into the assembly machine. It is die cut to the proper shape and size. This machine can assemble fuses with ten different amperage ratings. The fusible link material is fed into a punch-and-die station that has easily replaceable punch-and-die sets for each fuse amperage. A storage loop is built up in advance of a station, which allows the machine to be self-inventorying. That means that the punch-and-die station does not have to be synchronized with the rest of the machine.

Once the shape of the fusible link has been stamped, it is advanced to a placement station. The fuse body material is supplied in a continuous length from a large reel and has been preprocessed. The plastic body has been molded onto current-carrying links. As the fusible length material is made in with the fuse body, a sensor checks to make sure that body material is indeed entering the machine. A subsequent sensor verifies that a link has been placed.

The body material is aligned with the link. A very important design feature cuts and disposes of scrap zinc, keeping the machine run-

(continued)

ning smoothly and the workplace clean. Cams operate the placement tooling that assure the necessary synchronization of the carrier strip and the fuse body material. The fusible link is placed into the fuse body. The needles advance the assembly through a placement mechanism. Scrap is cut and dumped into a container.

The assembly then goes through two welding stations. These are commercially available spot resistance welders. A high current passes through the electrodes, creating a pool of molten metal, that resolidifies when the current stops. The fusible link must be welded in four locations. Each station puts two welds on one side of the fuse. Both sets of welds must be precisely located ⅜ in. from each other on the fuse. A bow in the carrier strip between the two welding stations permits the flexibility necessary to accommodate inherent length variations. Sensors are used to verify the presence of fusible body material in the station. They will inhibit the weld cycle and down-line operations if no material is in place.

At the next station, a plastic window is placed on the fuse to protect the fusible link. This window is die cut from a reel of plastic material. It must be placed precisely within close tolerances—eight-thousandths of an inch for the width and two-thousandths of an inch for the length. The window is inserted in the opening previously molded onto the fuse body. A sensor recognizes that the plastic material is there and will stop the machine if, for instance, the wheel runs out. Dancer rollers control the tension on the relatively delicate plastic material. The window is punched from the strip and advanced from the tooling to be placed in the opening of the fuse body. The window is fixed to the fuse at an ultrasonic welding station. Tooling contacts the plastic material and a high-speed vibration, 40 kilocycles per second, melts the plastic briefly, then it resolidifies, joining the window to the fuse. A tension wire at the center holds the tiny plastic window in place during ultrasonic welding operation.

The final steps take place at a vertical indexer station. As individual fuses are cut, the fuse blades are end cut to the proper form, a chiseled edge. Again, scrap is automatically disposed of and the fuse is inserted into a pocket on the eight-station indexer.

The fuse is tested to make sure that it contains all the right parts. If it passes the test, it is hot stamped with the proper amperage rating. If the zinc fusible strip is missing, the indexer will skip and no hot stamp will be applied. The incomplete fuses are segregated and put into a scrap bin. The completed satisfactory fuses are deposited in the collector bin, an important cost-saving part of the overall design of the machine.

The whole assembly process is supervised with a Texas Instruments programmable controller, located in the control cabinet. It can be programmed in three ways: (1) to run each function separately, (2) to run all functions through one complete cycle, or (3) to repetitively cycle all functions automatically. Sensors send information to the controller at each step of the assembly process. They check to make sure that material is being fed into the stations and they check to make sure the operation has actually occurred. The machine checks itself as the input/output lights show. Assembly integrity actually supervises the machine's operations.

An operator's panel has a visual display that can show the operator up to 24 different messages. It calls the operator's attention to such things as when the fusible link material has run out, or when the window material needs replacing. The result of this complex, fully automatic machine is a completed blade fuse produced inexpensively, automatically, and tested for quality, another application of automated assembly techniques.

VIBROMATIC

An important adjunct to the assembly operation is the parts feeders. These relatively inexpensive feeders often perform the same function as the expensive vision systems and robots that were developed in the early 1980s. Vibratory bowl feeders, for instance, create a bank of parts in a confined state, and in the proper orientation for an assembly operation.

At one manufacturer, Vibromatic, of Noblesville, Indiana, each bowl is designed individually. The engineers and craftsmen design the bowl and its tooling around the shape of the part to be fed and the speed at which it must be supplied. They take advantage of differences in profile and weight to help sort and orient the parts. A part to be sorted should be as nonsymmetrical as possible, or it should be symmetrical around its axis. Adding a tiny lip or flange to a part can help make the feeding process much simpler.

(continued)

Tom Terry, vice president of Vibromatic, explains how a feeder can sort the parts as they are fed into a system. According to Terry, from time to time, parts such as bolts and screws contain a variety of mixed-in foreign material in quantities that create problems in the actual production feeding systems, because the feeder is not capable of handling all of the variations. When that occurs, not only on screws, but on any part, oftentimes a trackability sorter for the particular part is built. Take, for example, a bolt sorter that checks for various dimensions and their integrity. At the beginning of the bowl, it checks across the flange of the head to ensure that the dimension is sufficient. If the bolt is oversize or undersize, it will fall through the parallel and go through a hole and into a catch track that will carry it to a bad parts tray. Good parts are carried across and dropped through a good parts opening, caught by a separate track, and tracked to be rehung. The parts are then checked for a roll thread with a parallel blade. A roll thread is larger than the blank, so any part that does not have any thread roll on it would hang on the parallel and be ejected as a bad part. Any part that has a thread would carry across the parallel and be returned to a good part exit by an airjet. Sorters of this type are not meant to be the only quality control of parts. However, they will make parts that can be fed well in a production condition.

Vibratory feeders and sorters are all custom made. They do, however, come in standard sizes. The largest size for a standard vibratory feeder bowl is 36 inches, and the smallest standard size is 6 inches. Bowls can be lined with a plastic brush material to help protect the integrity of parts. This type of liner is best used with parts that are dry.

Another type of feeder is the orbital bowl, designed for high-speed feeding of parts in the container industry. It can supply 2000 parts per minute to the customer's equipment.

Vibratory feeders need not be bowls; there are also straight-line vibratory feeders.

Straight-line feeders can handle much larger and heavier parts than bowl feeders; they are limited only by the size of a part a person can handle.

Feeders can be used in applications that are not entirely automated. For instance, a feeder may be an operator-assist machine that assembles two parts that are difficult to orient correctly, feeding them onto an assembly table for a final manual step. It replaces five or six operators with one to gain greater output, and parts are assembled correctly every time.

Such a feeder would have two sides. Ferrules are oriented and fed on one side and socket cap bolts on the other. Bolts are placed manually into the storage side of the feeder. The action of the feeder vibrates them up the incline and onto the orientation side. As they move along the tooling, the bolts are forced into proper orientation or dropped back into the feeder. As they move along the tooling, a "plow" drops the excess off, lining them up one at a time. Next, a scallop cut in the tooling eliminates the bolts that are upside down. Rejects are knocked down to the feeder again. In the last step, a "bubble" knocks extra bolts off the path so they don't build up and jam.

On the other side of the feeder, similar tooling sorts, orients, and positions the ferrules in a track just above the sockets. A switch senses when too many ferrules have built up, turning off that side of the machine until the loads have evened out.

As both parts feed through the final line, a probe reaches up, picking up both parts and putting them on a track. The track feeds parts onto the table, where an operator does the final assembly.

Another example of an operator-assist assembly is a feeder for solder rings. Rings are very delicate and must be handled gently. In addition, they are very difficult to load onto their storage rods by hand.

First, the rings are put on edge, in a standing position. Next, the rings are sorted for outside dimension and for flatness or warp. The good rings drop through an opening onto a rod on the outside of the feeder. The operator tends the rod, replacing it when full. The storage rods are capped and shipped. Such a feeder can handle several different sizes of rings simply by changing the tooling and the size of the mandrels and spear points.

Major Topics Covered

- Automated Guided Vehicle (AGV) Defined
- Historical Development of AGVs
- Rationale for Using AGVs
- Types of AGVs
- Guidance of AGVs
- AGV System Management
- Vehicle Dispatch
- AGV System Monitoring
- AGVs and Safety
- Trends in AGVs
- Other Materials Handling Methods
- *Case Study: General Motors Corporation*
- *Case Study: 3M Company—Magnetic Division*

Chapter Two

Automated Assembly

Automated Guided Vehicles

CAD/CAM

Numerical Control

Industrial Robots

Lasers in Manufacturing

Programmable Logic Controllers

Flexible Manufacturing

Computer Integrated Manufacturing

Other Related Technologies

AUTOMATED GUIDED VEHICLE (AGV) DEFINED

Materials handling is a broad term given to the various processes through which materials are transported, loaded onto machines, and unloaded from machines. In a manufacturing plant, materials needed to produce a product are ordered. Once received, the materials must be moved to a storage location. They are then moved from storage to production, where they are moved again from operation to operation. At each operation, the materials are presented to a machine, then they are unloaded and moved to a storage point for eventual shipment.

Moving materials from storage, through various production processes, and back to storage for shipping requires a great deal of materials handling. This can be accomplished using a number of different methods. Some of the more widely used materials handling methods and technologies are the following:

1. human handling
2. conveyor systems
3. fork-lift trucks
4. robots
5. pick-and-place units
6. automated guided vehicle systems

One advanced manufacturing technology that will see wide-scale use is the **automated guided vehicle (AGV)**. As the world of manufacturing continues to evolve toward the fully automated factory of the future, AGVs will play an increasingly important role. This chapter provides the fundamental knowledge manufacturing personnel need to understand the role AGVs will play in the factory of the future.

An AGV is a computer-controlled, driverless vehicle used for transporting materials from point to point in a manufacturing setting. They represent a major category of automated materials handling devices. An AGV can be used for any and all materials handling tasks from bringing in raw materials to moving finished products to the shipping dock.

In any discussion of AGVs, three key terms are heard frequently:

1. guide path
2. routing
3. traffic management

The term **guide path** refers to the actual path the AGV follows in making its rounds through a manufacturing plant. The guide path can be one of two types. The first and oldest type is the embedded wire guide path. With this type, which has been in existence for over 20 years, the AGV follows a path dictated by a wire that is contained within a path that runs under the shop floor. This is why the earliest AGVs were sometimes referred to as **wire-guided vehicles**. The more modern AGVs are guided by optical devices.

The term **routing** is also used frequently in association with AGVs. Routing has to do with the AGV's ability to make decisions that allow it to select the appropriate route as it moves across the shop floor. The final term, **traffic management**, means exactly the same thing on the shop floor that it means on the highway. Traffic management devices such as stop signs, yield signs, caution lights, and stop lights are used to control traffic in such a way as to prevent collisions and to optimize traffic flow and traffic patterns. This is also what traffic management means when used in the context of AGVs.

HISTORICAL DEVELOPMENT OF AGVs

Although AGVs are only now emerging as a widely used materials handling technology, they have actually been in use on a limited basis for over 30 years. The historical development of AGVs has paralleled that of other electronics-based technologies. The earliest AGVs relied on vacuum tube technology. As electronics technology advanced into the transistor stage, so did AGV technology. Just as computers and other related electronics technologies began their rapid emergence and wide-scale use with the development of programmable integrated circuit technology, so did the AGV. It is the use of the computer, coupled with optical devices, that has allowed the accelerated emergence of AGV technology to take place during the past decade.

At this time, AGVs have become an integral part of most heavily automated manufacturing plants. It is not uncommon for the AGV to be a principal materials handling technology in a modern automated manufacturing plant. The key to the wide-scale emergence of AGVs in modern manufacturing plants is the flexibility that computer control gives them.

RATIONALE FOR USING AGVs

Some manufacturing plants still use traditional materials handling systems. Some use automated storage and retrieval systems. Others use AGVs. Many use all of these together. Manufacturing technology students should understand why manufacturing firms use AGVs. Five of the most frequently stated reasons are summarized in Figure 2–1; they are as follows:

1. Because they can be computer controlled, AGVs represent a flexible approach to materials handling.

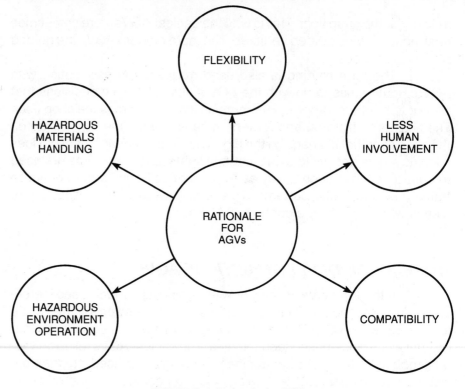

FIGURE 2-1 Rationale for AGVs.

2. AGVs decrease labor costs by decreasing the amount of human involvement in materials handling.
3. AGVs are compatible with production and storage equipment.
4. AGVs can operate in hazardous environments.
5. AGVs can handle and transport hazardous materials safely.

Flexibility is one of the keys to improved productivity in the modern manufacturing setting. It means the ability to adapt quickly to changes in products and processes brought about by the ever-changing demands of the marketplace. For a company to maximize its productivity and, thereby, its competitiveness, all manufacturing systems must be flexible. Computer control makes AGVs more flexible than traditional materials handling systems. Another factor that interests industrial engineers, manufacturing engineers, and manufacturing managers is the decreased amount of human involvement in materials handling that results with AGVs. Human involvement in materials handling adds significantly to the direct labor costs associated with manufactured products. It also increases indirect costs such as those associated with insurance and medical care necessary when human workers handle dangerous materials. By decreasing the amount of human involvement in materials handling, AGVs cut both direct and indirect manufacturing costs.

Another key feature of AGVs is that they are easily interfaced with existing production and storage equipment. Because they are so versatile, they can be adapted to be compatible with most production and storage equipment that might exist in a typical manufacturing setting. The AGVs also appeal to industrial engineers, manufacturing engineers, and manufacturing managers responsible for producing products in hazardous or special environment. The AGVs that are designed to operate in a hazardous environment can be as tough and durable as the environment demands and they can transport heavy loads. One of the most frequently stated reasons for using AGVs in the modern manufacturing plant is their ability to handle hazardous material. Occupational safety and health regulations coupled with a concern for human workers have made it nearly impossible to handle hazardous materials productively in a manufacturing setting.

Of the various reasons frequently given for using AGVs, perhaps the two that are the most important to the future of manufacturing are flexibility and compatibility. Their flexibility and compatibility allow AGVs to fit in with trends in the world of manufacturing, including automation and integration of manufacturing processes.

TYPES OF AGVs

Automated guided vehicles are called on for use in a variety of different manufacturing settings. Consequently, there is no one type that will meet the needs of every setting. Figures 2–2 and 2–3 show typical

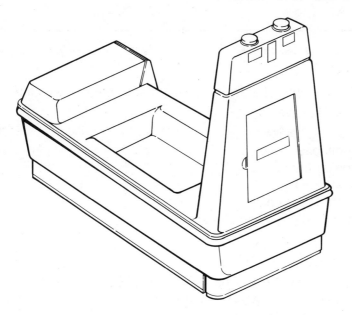

FIGURE 2–2 AGV and dock configuration.

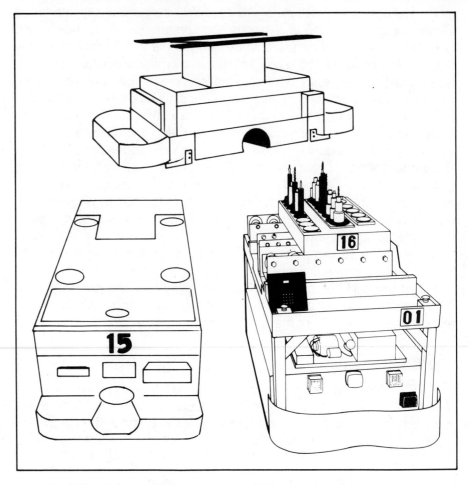

FIGURE 2–3 Sample of AGVs. *Courtesy of the Society of Manufacturing Engineers.*

AGVs. In the current state of development, there are six different types of AGVs (Figure 2–4):

1. towing vehicles
2. unit load vehicles
3. pallet trucks
4. fork trucks
5. light load vehicles
6. assembly line vehicles

Towing Vehicles

These AGVs are the work horses. **Towing vehicles** are the most widely used type of AGVs. Their most common use is in transporting

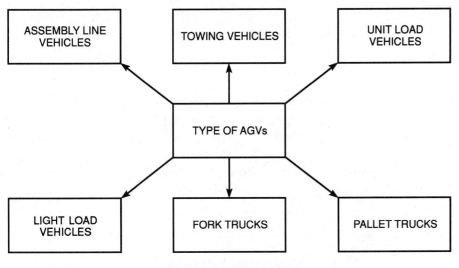

FIGURE 2–4 Types of AGVs.

large amounts of bulky and heavy materials from the warehouse to various locations in the manufacturing plant. A popular approach is to arrange a series of vehicles into a train configuration (Figure 2–5). In such a configuration, each vehicle can be loaded with material for a specified location and the train can be programmed to move throughout the manufacturing facility, stopping at each location. This type of AGV varies widely in size and capacity. Some towing vehicles can carry as much as 50,000 pounds of material or up to ten full pallets.

Unit Load Vehicles

Unit load vehicles represent the opposite extreme from towing vehicles. Whereas towing vehicles are used in settings requiring the movement of large amounts of material to a variety of different locations,

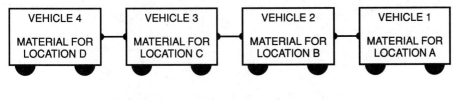

WEIGHT CAPACITY PER VEHICLE = 50,000 POUNDS
LOAD CAPACITY PER VEHICLE = 10 PALLETS

FIGURE 2–5 Towing vehicle train configuration.

unit load vehicles are used in settings with short guide paths, high volume, and a need for independent movement and versatility (Figure 2–6). Warehouses and distribution centers are the most likely settings for unit load vehicles. While towing vehicles might carry as many as ten pallets per vehicle, unit load vehicles handle individual pallets. Towing vehicles might be chained together and must, therefore, have all their movements coordinated. Unit load vehicles, on the other hand, operate independently of each other and might even pass back and forth by one another during operation. An advantage of unit load vehicles is that they can operate in an environment where there is not much room and movement is restricted.

Pallet Trucks

The **pallet truck** is different from other AGVs in that it can be operated manually. Pallet trucks are used most frequently for materials handling and distribution systems. They are driven along a guide path from

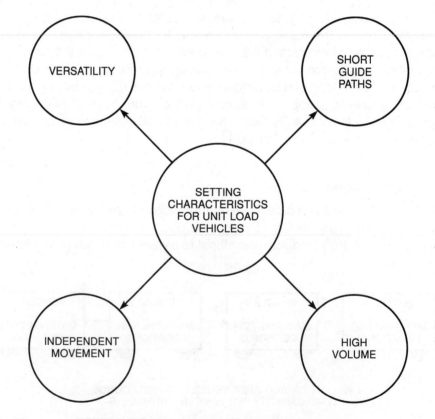

FIGURE 2–6 Setting characteristics for unit load vehicles.

location to location and are unloaded as they go. Because they can be operated manually, pallet trucks represent a very flexible approach to materials handling.

Fork Trucks

The **fork truck** type AGV is to the automated manufacturing plant what the fork lift is in a traditional materials handling setting. Fork trucks are designed for use in highly automated manufacturing plants. They are used when it is necessary to pick material up at the shop floor level and move it to a location at a higher level or to pick up material at a higher level and move it down to the shop floor level. Unlike the traditional fork lift, however, fork truck type AGVs travel along a guide path.

Light Load Vehicles

Light load vehicle technology is simply the miniaturization of unit load vehicle technology. Light load vehicles, as the name suggests, are used in manufacturing settings where the material to be moved is neither heavy nor bulky. The typical light load vehicle is limited to a weight capacity of 200 pounds or less. The most common application of the light load vehicle AGV is in electronics manufacturing settings. For example, a light load vehicle might transport small containers of electronic components from a storage area to the assembly area.

Assembly Line Vehicles

As the name implies, **assembly line vehicle** type AGVs are used in conjunction with an assembly line process. Their most common use is in the assembly of automobiles. Assembly line vehicles can be used to transport major subassemblies such as automobile engines, transmissions, doors, and other associated subassemblies to the proper location on an assembly line. Using such vehicles can enhance the flexibility of an automobile assembly line.

For example, assembly line vehicles can be programmed to transport subassemblies to specified points along the assembly line in order. They can also be programmed to bypass certain locations on the assembly line and come back to them. This becomes a particularly valuable capability when a certain part of the assembly line breaks down. By reprogramming the assembly line vehicle, it can be made to bypass the inoperative portion of the assembly line and go on to other areas along the line that are still in operation.

GUIDANCE OF AGVs

Guidance of AGVs is a key issue in the ongoing development of this technology. The guidance hierarchy of AGVs is represented by the pyramid in Figure 2–7. At the bottom of the pyramid is manual guidance of AGVs. The next step in the pyramid is computer and optical guidance of AGVs. This is the current state of the art. The highest level on the pyramid is self-guidance. Self-guidance represents the optimum in AGV technology. Modern AGVs that rely on optical guidance systems represent an important step forward from both manual and wire guidance systems.

However, even optical guidance systems have their shortcomings. Optically guided AGVs rely on sensing devices that follow a highly reflective line along the shop floor. The line may be painted on the floor or a special reflective tape may be used. In either case, the line can be obstructed by dirt, debris, or normal wear. In such cases, optical guidance systems become unreliable. When self-guidance of AGVs is accomplished, this technology will have achieved its fullest potential.

There are two approaches used for steering AGVs:

1. differential speed steer control
2. steered-wheel steer control

An easy way to understand the difference between differential speed steer control and steered wheel steer control is to think of the former as the type of steering used in a military tank and the latter as the type of steering used in an automobile (Figure 2–8). An AGV with a differential speed steer control system uses two independent fixed wheel drives on either side of the vehicle. This allows the drive on one side to operate at a faster speed than the drive on the other side, thereby causing the vehicle to turn. The **amplitude detection guidance sensor** feeds

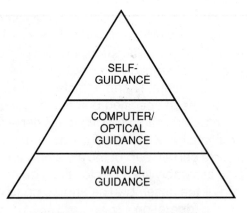

FIGURE 2-7 AGV guidance systems.

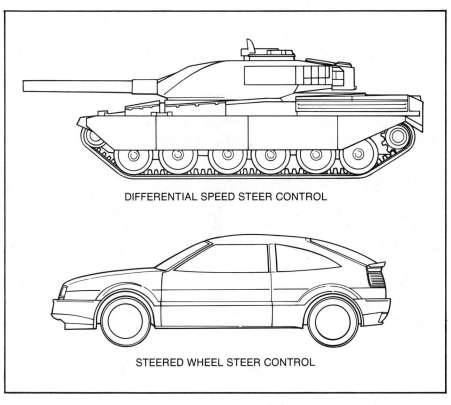

FIGURE 2–8 AGV steering systems.

required data to the drive systems so that they know when to turn. An AGV using the differential speed steer control system has left and right sensing devices mounted on the front of the vehicle. These sensors receive information for left or right turns. The amplitude detection guidance sensor balances the signals received from the left and right sensing devices.

For example, if one sensing device sends a signal that is greater in amplitude than that of the other sensing device, the AGV turns accordingly until the signals are once again balanced between the sensors. The sensors actually receive the information by sensing the reflection from the paint strip or tape strip that marks the guide path on the shop floor.

The steered-wheel steer control system is easier for most people to understand because it operates in a manner similar to that of an automobile. The AGVs that use this type of steering system have a single front wheel that rides along the paint or tape strip that marks the guide path along the shop floor. A phase detection guidance sensor detects whether the wheel is on course, steering to the left of the guide path, or steering to the right of the guide path. As the phase detection guidance sensor sends this data back to the AGV, the wheel at the front of the AGV turns to compensate so that the AGV is in phase.

Steered-wheel steer control systems can be used with assembly line vehicles, towing vehicles, unit load vehicles, pallet trucks, fork trucks, and light load vehicles. Differential speed steer control systems, however, are limited to use with assembly line vehicles, unit load vehicles, fork trucks, and light load vehicles. They are not used with towing vehicles or AGVs that can be manually operated, such as pallet trucks.

AGV SYSTEM MANAGEMENT

In any manufacturing plant employing AGVs as the principal means of automated materials handling, there is a need to manage the systems. There are three key issues in AGV system management (Figure 2–9):

1. traffic control
2. vehicle dispatch
3. system monitoring

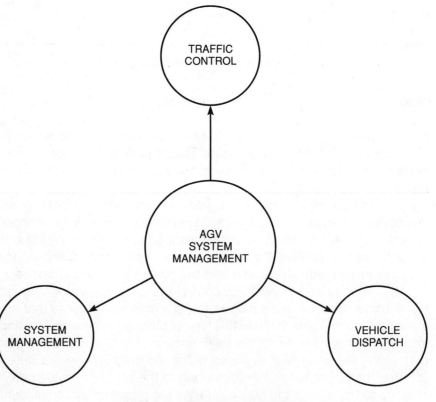

FIGURE 2–9 AGV system management.

Traffic Control

Manufacturing plants that use AGVs as the principal method of materials handling need a traffic control system. There are three types of traffic control systems used with AGVs (Figure 2–10):

1. zone control
2. forward sensing control
3. combination control

The **zone control system** is the most widely used. With this system, the guide path areas of the shop floor are divided into zones (Figure 2–11). The underlying premise of this control system is that only one AGV is allowed in a given zone at a time. Figure 2–11, shows two towing vehicle AGVs: vehicle A and vehicle B. Vehicle A is stopped in zone 2 and vehicle B is stopped in zone 1. Note that zone 2 has three unloading stops and zone 1 has two unloading stops. As vehicle A moves forward to each stop in zone 2, vehicle B must maintain its position. If vehicle B moves into zone 2 when vehicle A moves forward to stop 3 of zone 2, it would violate the underlying premise that only one vehicle can be in a specified

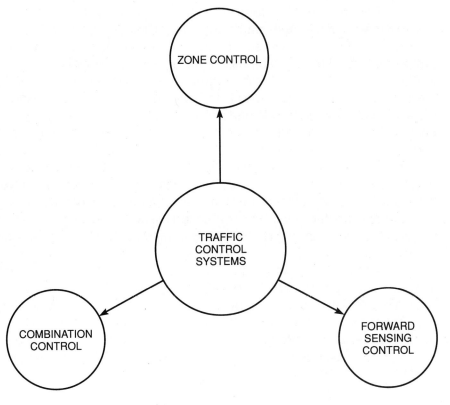

FIGURE 2–10 Traffic control systems.

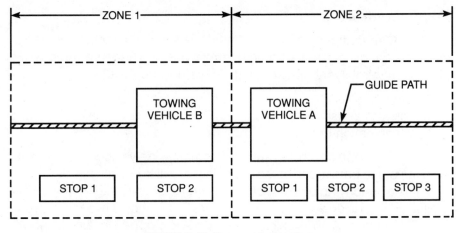

FIGURE 2–11 Zone control system.

zone at a time. Consequently, with a zone control system, vehicle B cannot move into zone 2 until vehicle A clears that zone. This AGV traffic control system can be used with straight, curved, and combination guide paths.

Forward sensing control is a traffic control system that works well when most of the guide paths are straight. However, this traffic control system does not work effectively in situations involving curved guide paths or intersecting guide paths. Forward sensing control is illustrated in Figure 2–12. With this traffic control system, each AGV has one or more sensing devices mounted on its front. There is a specified range or distance that must be maintained between AGVs. The sensing devices continuously "look" forward to sense any object that falls within the specified range. If one vehicle gets too close to another vehicle, the sensors will detect that vehicle and cause the one in the rear to stop. In a chain of vehicles such as that in Figure 2–12, vehicle 2 can only move when vehicle 1 is outside and forward of the specified range. Forward sensing control does offer the advantage of allowing the specified range between vehicles to be very short. In this way, a greater density of AGV traffic can be accomplished than with zone control.

Although it is not frequently done, zone control and forward sensing control can both be used within the same traffic control system. Such a **combination control** approach is called for when the guide path system contains long stretches of straight guide path but still has intermittent curves and/or intersections. When this is the case, forward sensing control can be used for traffic control on the straight portions of the guide path and zone control at the curved and/or intersecting areas along the guide path.

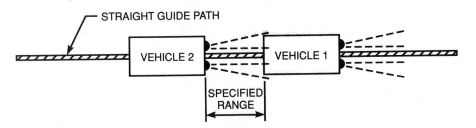

FIGURE 2–12 Forward sensing control.

VEHICLE DISPATCH

Since AGVs are automated instead of having a human operator on board, there must be some means for dispatching AGVs along the guide path routes. There are five methods used for dispatching a AGVs (Figure 2–13):

1. onboard dispatch
2. offboard call systems
3. remote terminal
4. central terminal
5. combination dispatch

The most widely used method of vehicle dispatch is the on-board dispatch approach. With this method, each AGV has a control panel mounted on it. A human operator uses this control panel to pro-gram the AGV stops along the guide path. The control panel can also be used to program the AGV activities at each stop. This is what makes on-board dispatch so flexible. As an AGV moves from stop to stop along the guide path, it may need to perform a different function at each stop. With onboard dispatch, the human operator is able to program these func-tions into the AGV using the onboard control panel.

In situations where the process of transferring materials from the AGV to a stop location is automated, the offboard call system method of vehicle dispatch is effective. Offboard call systems range from very simple to complex. The simplest form of offboard call system is a single button at a given station along the guide path that a human operator may press to call an AGV to that station or to stop it at that sta-tion as it passes by. At the other end of the spectrum is the more sophisti-cated offboard call system in which a complex panel or call box can be used to call a vehicle to a specified location and dispatch it to one or more other locations along the guide path.

Onboard dispatch and offboard call system dispatch are de-centralized approaches to vehicle dispatch. The trend is toward more

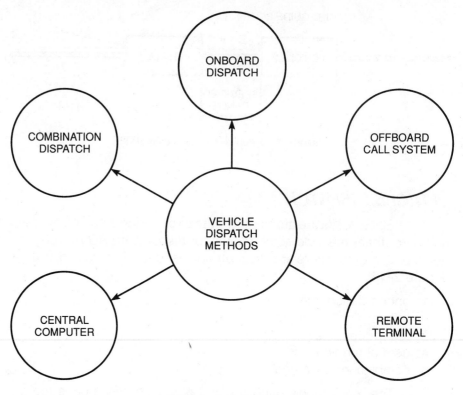

FIGURE 2-13 Vehicle dispatch.

centralized control, the ideal being all AGVs dispatched from one location. A move in that direction is the remote terminal method of dispatch. With this method, a human operator at a remote central location has a diagram of the shop floor displayed on a computer terminal. Symbols representing individual AGVs tracking of the movement of individual vehicles within the overall system. Using the computer keyboard, the human dispatcher is able to move AGVs throughout the system from one central location.

Note that although a computer terminal is used as a tool in dispatching AGVs with this method, remote terminal dispatch is not computerized dispatch. With this method, a human operator still moves the AGVs around the shop floor, using the computer as an aid in doing so. This approach might be called manual control of the AGV.

Computer vehicle dispatch is called central computer dispatch. This is the highest level dispatch method within the hierarchy of dispatch methods. With such a system, the AGVs are programmed by computer to move along the guide path and make the necessary stops. They are also programmed for functions to be performed at each stop. Once the program has been written and the instructions from the program are issued to the AGV, no further human interaction is necessary. It is not necessary for a human operator to sit in front of a terminal and visually monitor the movement of the AGVs as it is with the remote terminal dispatch method. As manufacturing moves closer and closer to full automation, this vehicle dispatch method will become the norm.

Although it is not frequently done, it is possible to combine one or more of the various vehicle dispatch methods. For example, a manufacturing firm using central computer dispatch of its AGVs might also have a remote terminal, offboard call system, or onboard dispatch system available. This secondary system would be used when the primary system malfunctioned or was temporarily inoperative. Whether or not combination vehicle dispatch methods can be used and the methods that can be combined depends on the design of the individual AGV system that the vendors have built into the system or are willing to build into the system for a given application.

AGV SYSTEM MONITORING

One of the shortcomings of an automated materials handling system that relies principally on AGVs is that a malfunction of one vehicle can cause problems throughout the entire system. Therefore, it is important that AGV systems be monitored closely and continually. The ideal monitoring system is one that gives human operators instant, real-time feedback on the following:

1. location of all vehicles within the system
2. location of malfunctioning or inoperative vehicles
3. movement of vehicles
4. amount of time vehicles spend at each stop and en route between stops
5. status of all vehicles in the system, loaded or unloaded
6. destination of all vehicles within the system
7. status of the batteries in all vehicles within the system: charged, charging, or weak

The sophistication of the monitoring system is dictated by the sophistication of the overall AGV system. For example, AGV systems that rely on simple dispatch methods such as onboard dispatch or offboard call system dispatch require little monitoring. This is because the human operators who use the onboard dispatch panel or offboard call box visually monitor routinely. However, AGV systems that use sophisticated central computer dispatch methods require very close and continual monitoring because even the slightest programming error or bug in a program can cause problems throughout the entire AGV system.

There are three types of systems widely used for monitoring AGV systems (Figure 2–14):

1. locator panel monitoring
2. computer display monitoring
3. performance report monitoring

Locator Panel Monitoring

This is the simplest, least sophisticated type of AGV system monitoring. The panel is a series of lights within zones on the panel that correspond with zones on the shop floor. When an AGV enters a given zone, the light on the panel corresponding to that zone illuminates. When the AGV moves out of that zone, the light goes out and the light for the

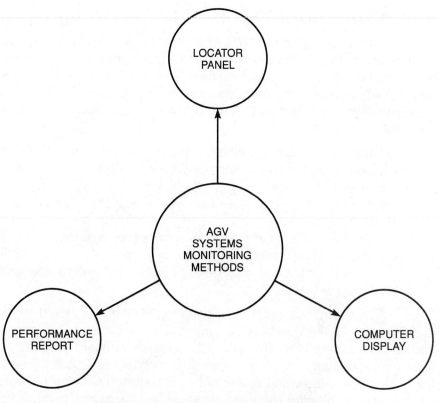

FIGURE 2–14 AGV systems monitoring methods.

next zone illuminates. This particular monitoring method tells the operator only that an AGV is in a given zone. It does not tell which AGV, whether the vehicle is loaded or unloaded, what the vehicle's destination is, or whether the vehicle's battery is charged or getting weak. Some locator panels are equipped with timers so that the human operator can monitor how long a light on the panel stays illuminated so that the operator is able to tell if a vehicle stays in a zone longer than it should. This would indicate a breakdown or malfunction of some type. However, it does not tell the operator what the problem is or even which vehicle is malfunctioning.

Computer Display

The most sophisticated type of AGV monitoring system is the computer display system. Monitoring data can be displayed on a computer terminal in two forms: graphic or alphanumeric. With the graphic approach, the human operator has a graphic representation of the AGV system that shows all vehicles, their locations in real time, the locations of any malfunctioning vehicles, movement of vehicles through the system, amount of time vehicles spend between stops, whether a vehicle is loaded or unloaded, destinations of all vehicles, and the status of each vehicle's battery. This monitoring method gives the human operator all the information needed to take necessary action when a malfunction or breakdown occurs.

The alphanumeric approach to monitoring AGV systems on a computer display lists all of the same information, but in table form that is updated continually. Most subsystems have a built-in alarm that causes the data representing a given vehicle to flash or illuminate when a malfunction occurs. In addition, a description of the malfunction is displayed on the terminal.

Performance Reports

Performance reports are logged records of the type of information displayed on the computer terminal with computer display monitoring of AGV systems. Performance reports give the historical record of the performance of each AGV within the system. For example, by examining the performance report of an AGV system for the immediate past month, an operator can determine how much time vehicle X spent at every stop within the system every day in the month. Such information is valuable to industrial engineers responsible for continually improving the efficiency of an AGV system.

AGVS AND SAFETY[1]

Safety in automated guided vehicle systems is a characteristic feature that must be designed into the system prior to installation. Safety, like quality, cannot be an afterthought. All AGV systems interact with operators and other personnel in varying degrees. Provisions must be established early in the design phases not only to safeguard the transportation of materials, but to reasonably protect personnel in the immediate area of the vehicle system.

There are two main objectives to safety devices in AGV systems. First, the devices prevent contact or collisions between vehicles and personnel or other vehicles. Second, in the event of malfunction of the anticollision operation, the primary safety devices would immediately render the vehicle inoperative until manual intervention occurred to rectify the situation.

It is important to note that with ever more complex system requirements, safety devices must be provided in initial system and equipment formulations. This section focuses on the hardware and software controls and devices. Application parameters and characteristic features of each are described.

Safety Device Classifications

The most straightforward classification segmentation of safety devices is the differentiation between fixed and mobile equipment. Fixed equipment generally consists of system-level devices. These include blocking systems and general layout considerations of facilities. Mobile equipment has considerations that must be observed by both the manufacturer and the user. It is the user's responsibility to ensure that the vehicle system is properly used. This includes proper operation and maintenance training as well as operation within design parameters. A frequent violation of design parameters is in load stability where the dimensions or center of gravity change with product variation.

The manufacturer has many more parameters for concern. Monitors must be provided for guidance and speed, and provisions must be incorporated into the vehicle for motive power disconnect in the event of malfunction. Operator protection must be provided in the form of guards and deadman controls for manual operation. Additional safety devices

---

[1]This section was provided courtesy of the Society of Manufacturing Engineers from a paper presented at SME's Automated Guided Vehicle Conference, 1986, by Robert R. Lasecki, Vice President of Austin Consulting.

may be requested by the user in the form of look-ahead, noncontact, collision avoidance sensors and special lights such as turn signals.

Blocking Systems

There are a number of systems currently in use to control vehicle traffic and prevent collisions. These systems range in complexity from hardware-intensive in the facility to software-intensive on the vehicles with many levels in between. Current systems or techniques in use include the following.

Bumper Blocking

This is the simplest of all implementations and is often applied in smaller vehicles in assembly line operations. Bumper blocking provides high-density vehicle queuing in a small space. Vehicles stop on contact with the front bumper. Vehicle speed is typically restricted and travel distances are relatively short. After contact with the bumper, a vehicle will remain stopped until after the bumper activation force has been removed (vehicle ahead moves) plus a preset time delay. Typical delays range from five to thirty seconds.

In-Floor Zone Blocking

This is the oldest form of zone blocking applied to automated vehicles. The principles are those used by the railroad industry for over a century. The guide path is physically divided into logical zones. Vehicles are typically permitted to enter a zone only when the next zone is completely available. The in-floor implementation uses sensors in the floor at zone boundaries to detect vehicles passing. The traffic control logic is external to the vehicles and wired to the floor. Vehicles are released for travel into a zone by means of an enable beacon or, even more simply, by having the path energized.

With this system, the vehicles have no need to monitor traffic conditions and can exist with minimal control logic. Modern implementations using this traffic control technique use programmable logic controllers to provide the floor control logic instead of hardwired relay logic. The removal or addition of a vehicle in this type of installation requires manual intervention with the control logic.

Computer Zone Blocking

This is a technique of traffic control similar to the in-floor zone technique except it allows more complex traffic situations. All monitoring of vehicle positions is done by a central minicomputer. Vehicle positions along the guide path can be externally monitored or communicated directly from the vehicle either through the guide path or by radio.

The use of bumper blocking is the only contact method of traffic control for AGVS. This technique, when also used as the primary safety mechanism, must be designed in a totally fail-safe manner. The only permissible connections are direct to motive power disconnect as specified in the American National Standards Institute (ANSI) B56.5 code. The use of interim logic, microprocessor software, or other signal conditioning equipment is specifically disallowed. No single point failure should render the vehicle unsafe. Computer and hardware-based in-floor blocking techniques are noncontact. In normal operations, they maintain vehicle separation of a minimum design distance. In these systems, the bumper becomes the primary safety in the event of system failure and impending vehicle or load contact.

General Layout

There are many parameters within facilities that affect the layout of an AGV system. All areas involving pedestrian travel or interaction with workers require a minimum clearance of 18 in. (46 cm) between a vehicle and its transported load and any fixed object. This clearance distance allows an average person to stand and have a vehicle pass without presenting a safety hazard. Specific areas where clearance cannot be maintained, such as at load transfer stations, need to be designated as restricted areas. These areas are designated for vehicle use only and all pedestrian traffic is prohibited. Such areas are typically designated by having signs posted and the floor striped.

Special fire door controls are required to clear vehicle traffic to allow the doors to close in the event of a fire as well as to prevent vehicle traffic from attempting to cross a closed fire door. The most common technique is to use a limit switch on the door mechanism to energize the guide path in the immediate vicinity of the door only if the door is in the fully open position.

Pedestrian traffic should be routed in areas not traveled by the AGVs. Locations where vehicle traffic crosses the pedestrian aisles require signs and appropriate alarms. Lights indicating an approaching vehicle are most common but horns and bells are also used. Similarly, warning indicators should be provided on all doors that can be automatically activated by vehicle traffic to avoid personal injury.

A critical systems issue with respect to AGV systems is floor condition. Vehicle ratings are determined and specified on a level dry surface with a particular coefficient of friction. Floors that are poorly maintained and become rough will subject vehicles to excessive vibration and induced failures. Floors that have oil, grease, or other residue buildup can become slippery due to a degraded coefficient of friction. This can easily result in diminished steering accuracy due to lateral slipping as well as poor stopping accuracy.

Under these conditions, emergency stop actions become quite dangerous as there is minimal tractive force available to stop the vehicle. The result could easily be serious impact between the obstacle (such as a person) and the load or the vehicle frame.

Standing water on floor surfaces is a common cause of AGV accidents. In a wet environment, the tractive force of the vehicle generally becomes unpredictable, and bumper range and stopping distances are no longer anywhere near the vehicle design values. As a result, experienced vehicle manufacturers include equipment disclaimers regarding wet floors.

Bumpers

Bumper design, activation, and requirements have been some of the most controversial issues in the AGV industry for more than two decades. The primary purpose of bumpers is to detect an impending collision between a vehicle and a person, another vehicle, or obstacle, and to initiate an action that will stop the vehicle prior to contact with the vehicle frame. Bumpers are, therefore, typically extended about 18 in. (46 cm) in front of the vehicles. Additional bumpers in the forms of bat wings, side pressure switches, or whisker switches may be required to protect oversize loads.

The primary controversy is whether vehicles should be allowed to automatically restart after a bumper stop. In bumper blocking systems, manual restart requirements are totally counterproductive. In systems where the bumper is the primary safety, manual intervention following a bumper stop is required. Bumpers must, in all cases, be active in automatic modes of operation and provide protection in the direction of travel.

The ANSI B56.5 code requires that bumpers directly control motive power interrupts and do not require either software or hardware logic or signal conditioning (e.g., amplifiers, gates, or inverters) for operation. Bumper activation signals may, however, be monitored by processing circuitry for diagnostic purposes.

A special case, and a very difficult one for manufacturers to address, is the inclusion of an effective bumper on fork-type vehicles that are traveling in the fork direction. GMA pallets do not allow for the thickness of European type forks. Embedded optical sensors are neither failsafe nor reliably sensitive to provide appropriate protection to personnel and equipment.

Operation Monitors

There are various levels of operation monitors that are either required or recommended for vehicle system safety that affect vehicle manufacture. The two primary monitors generally required are for guidance and velocity functions. The ANSI B56.5 code requires loss of guidance in the automatic mode to result in an immediate stop of motive effort. This is independent of whether the guidance mechanism is optic, electromagnetic, or computed as in the case of an inertial guidance system. The loss of path detection technique is dependent on vendors and not all implementations are fail-safe. A fail-safe loss of path detection will render the vehicle motionless in the event of either loss of signal or failure of the circuitry.

In variable speed vehicles, overspeed detection is required. This function is especially critical in applications involving ramp operations. Overspeed conditions while traveling down ramps or grades can be extremely hazardous if not properly detected and corrective action applied. Overspeed detection is typically performed by measurement of drive motor terminal voltage, tachometer voltage, or distance measuring equipment pulse rate. All three techniques of detection are comparably reliable and are determined by the complexity of the control system.

Vehicles with automated load transfer capability require a handshake safety interface with the load stand to ensure that

- the vehicle is properly aligned for a safe transfer
- the mechanism is ready to receive the load and
- the load has not jammed before transfer is completed

These same vehicles may also have load detection sensors that perform dual functions. First, they indicate when a load has fully transferred onto the vehicle. Second, they can indicate when a load has shifted to an unsafe position during transport and activate the appropriate vehicle controls.

Vehicles with onboard intelligence in the form of one or more microprocessors should have a software lockup detection circuit. This circuit is typically referred to as a watchdog timer. It requires a software-timed trigger pulse within a specific interval. If the circuit does not receive the trigger pulse, it declares a fault condition that renders the vehicle motionless. Loss of the trigger pulse can be software based, but in fully commissioned systems it is generally caused by computer hardware failures.

Vehicles that can be operated manually, as well as in full automatic mode, are required to have deadman sensors. This type of sensor requires intentional activation by the operator for the motive power to be active. Typical implementation is through a spring-loaded foot switch for a rider vehicle or a hand switch on a pendant controlled vehicle.

Guards and Warning Devices

All moving portions of automated guided vehicles should be guarded to prevent injury from access to pinch points. This includes items such as exposed wheels, shuttle mechanisms, and lift tables. Combinations of audible and/or visual alarms are required on the vehicles for different functions. An alarm indication is required prior to initiation of vehicle movement to alert personnel in the area that motion is forthcoming. Abrupt changes in direction such as reversal for station entry or pivot steer operations require activation of an alarm for operations personnel safety. Audible alarms are typically electronic beepers for continuous annunciation and load electric horns for emergency-type alarms. Some vehicles also incorporate mechanical wheel bells to indicate vehicle motion.

Visual alarms are just as varied. Vehicle status (in automatic) and motion are typically indicated with either a strobe, rotating, or flashing light mounted in a location where it is highly visible from the periphery of the vehicle. Some users require the vehicles to be equipped with automatic turn signals to alert pedestrian traffic of route selection at intersections or path divergences.

Labels and signs can also be considered as safety items. Signs such as NO RIDERS and NO RAMP OPERATION for vehicles and AUTOMATED VEHICLE CROSSING for pedestrian aisles can contribute considerably to the safety of an installation. Battery charging and changing areas and vehicles designated for operation in hazardous environments also require special markings as designated by the Underwriters Laboratory and NFPA. Vehicle base color should be such that it renders the vehicle highly visible in all application environments.

Noncontact Sensing

Several devices are available that provide noncontact sensing of objects or obstacles in the path of a vehicle. These sensing methods are typically used to slow the vehicle down prior to collision and, as an alternative, may be used to stop the vehicle. The two most common technologies used for noncontact sensing are infrared optics and ultrasonics.

Optic Sensing

Optic sensing uses the reflective characteristics of objects for detection. An optical beam of infrared light is emitted forward from the vehicle. Objects in the path of the vehicle reflect a portion of the light to create the detection signal. This technique is sometimes used for vehicle noncontact separation control by placing a reflective target on the rear of each vehicle. This works well on straight aisles, but loses reliability on curves.

Some optic units can be momentarily blinded by sudden exposure to strong sunlight. Most will be activated by highly reflective surfaces farther than the normal activation range. It is typical to require some surfaces in the facility that pose problems to be painted with a nonreflecting paint. Also, some colors and textures of clothing will not be reliably detected by optic systems.

Ultrasonic Sensing

Ultrasonic sensing is a technique implemented by emitting a burst of ultrasonic energy forward from the vehicle and measuring the time from emission to reception of a reflected signal. Since acoustical energy travels at about 1,000 ft. (305 m)/second and the distances sensed are typically under 20 ft. (6 m), the pulse rate can be high enough to approximate continuous sensing. The received signal amplitude is sensitive to a number of system and target variables. The size and shape of the target, the acoustic absorption of the target, and the distance from the vehicle all affect signal strength.

A flat object at an acute angle to the path may never be seen. A thermal shear, as caused by an air curtain or an air conditioning vent, can appear as solid as a facility wall when viewed by ultrasonics.

Ambient noise in the upper acoustic spectrum can blur the reflected signal just as induced noise through vehicle vibration can generate false signals. Although not utilized by all vendors, ultrasonic sensing has one unique advantage over optical sensing in that it can be made reasonably fail-safe. Since there is a measurable time delay between the transmitted pulse and the first allowable return pulse, the transmitted pulse can be "heard" to determine proper transmitter operation. This is impractical to accomplish with optics.

User Considerations

There is a significant responsibility for the safety of an AGV system by the user. System success and safety are both predicted on acceptance of the system by the operations staff. Personnel that have not been properly exposed to the system or properly trained to interact with the system during normal operation can be subjected to hazardous situations. It is the responsibility of the user to see that new staff are appropriately trained and that operating rules and disciplines are enforced. Maintenance programs, including preventive maintenance, should be established and monitored.

A critical parameter often overlooked is load stability. Many facilities have a changing product, changing packaging techniques, or changing transport volumes. Any of these changes can result in a shifted center of gravity (CG) of the load. Vehicle dynamics are specified with a predetermined CG and shifted values can render the load unstable (during travel on curves or E-stop conditions) or cause difficulties in vehicle tracking or load transfers. With all major product or process changes, the CG of the load should be checked against the vehicle system design parameters. The vehicle system parameters can then be adjusted if required.

It is the responsibility of the user to maintain the guide path areas free from obstructions as well as clean and dry. Dirt, water, and grease on floors can be just as hazardous as loose boxes or a parked vehicle in the guide path area.

Conclusion

There are multiple types and categories of devices that are available for implementation in AGV systems to address safety concerns. System-level devices address general facility issues, as well as vehicle traffic management. Devices for installation on mobile equipment provide both primary and supplemental features to enhance safe operation of AGV systems. There are no better safety devices available than the people involved with the system. It is critical that safety be a primary issue during the design and implementation phases of development. Training and maintenance programs are imperative for continued safe operation of AGV systems.

TRENDS IN AGVs

AGVs have been in use for over 30 years. Their development has coincided with the development of such basic electronics technologies as the vacuum tube, transistor, and programmable integrated circuit. As manufacturing systems continue to evolve and become more and more automated, AGVs will become more important components of such systems. The more AGVs are used, the greater will be the bank of knowledge on their effective and efficient use. As users and vendors continue to interact in this regard, there will be continual improvements in AGV technology. Glen A. Graham in the Society of Manufacturing Engineers publication *Automation Encyclopedia* lists several trends to expect in AGV technology:

- Increased diagnostics
- Smaller packaging of hardware on the vehicle
- Faster vehicle-to-system communication
- Off-wire guidance
- Application of larger and more hardware-redundant systems at the system management and host levels
- Ability to download a new "route map" and path optimization formation to the vehicle automatically
- Increased use of horizontal interfaces to programmable logic controllers or other microprocessor control devices
- Use of complex vehicle task optimization and prioritization
- Improve monitoring capabilities of the system for the vehicles as well as for individual components
- Use of computer emulators to pretest the traffic and system management software prior to field installation

OTHER MATERIALS HANDLING METHODS

At the beginning of this chapter, the most widely used materials handling methods were listed. The list included human handling, conveyor systems, fork-lift trucks, robots, pick-and-place units, and AGVs. AGVs are covered in the previous sections. Of the remaining, the most widely used methods are human handling and pick-and-place units. Human handling is a manual method. Pick-and-place handling is an automated method. In some cases, human handling is more appropriate. In others, automated handling is better.

Human Handling

Human handling of materials involves the physical movement of materials using either the hands or with the aid of a nonautomated mechanical device. In deciding whether human handling is the optimum materials handling method for a given manufacturing setting, several criteria should be considered:

1. workpiece configuration
2. process time requirements
3. output volumes
4. operator skill
5. environmental health and safety

Human handling is appropriate for bulky, complex materials where there may be more handling than process time. Short-run, low-volume settings requiring a high level of operator skills are also appropriate for human materials handling, as are relatively safe, healthy settings.

Pick-and-Place Handling

When human handling is not appropriate, one method that can be used is pick-and-place handling. Pick-and-place units are the simplest form of automated materials handling manipulator. Such units are either pneumatically powered or cam controlled.

Pick-and-place units have such capabilities as a lift-and-lower capability, a reach-and-retract capability, wrist motion, body rotation, a grip-and-let-go capability, and vertical pitch. Figure 2–15 is an example of a pneumatically powered pick-and-place unit and Figure 2–16 is an example of a cam-controlled pick-and-place unit.

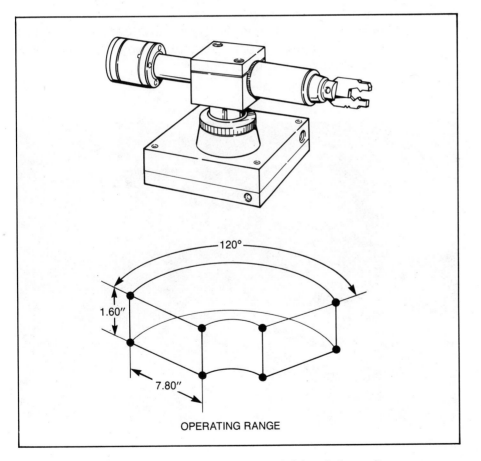

120°

1.60"

7.80"

OPERATING RANGE

FIGURE 2–15 Pneumatically powered pick-and-place unit.

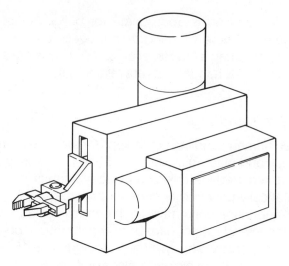

FIGURE 2–16 Cam-driven pick-and-place unit.

Figures 2–17, 2–18, and 2–19 are examples of materials handling/assembly systems with parts feeders, conveyor systems, pick-and-place units, and assembly machines.

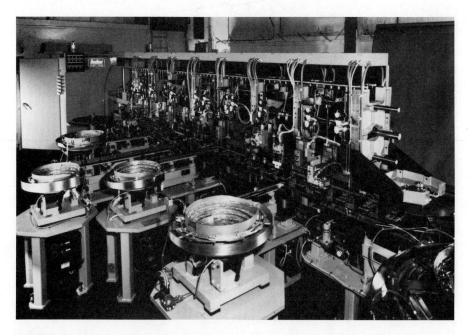

FIGURE 2–17 Materials handling and assembly system. *Courtesy of the Bodine Corporation.*

FIGURE 2–18 **Materials handling and assembly system for trigger pump heads.** *Courtesy of the Bodine Corporation.*

FIGURE 2–19 **Materials handling and assembly systems for videocassettes.** *Courtesy of the Bodine Corporation.*

KEY TERMS

Automated guided vehicle (AGV)
Guide path
Wire-guided vehicle
Routing
Traffic management
Towing vehicle
Unit load vehicle
Pallet truck

Fork truck
Light load vehicle
Assembly line vehicle
Amplitude detection guidance
    sensor
Zone control system
Forward sensing control
Combination control

Chapter Two REVIEW

1. What is an AGV?
2. Explain the following terms:
   a. Guide path
   b. Routing
   c. Traffic management
3. What is a wire-guided vehicle?
4. To what can the accelerated emergence of AGV technology in recent years be attributed?
5. Give five reasons why a manufacturer might use AGVs for materials handling.
6. List and explain at least three types of AGV.
7. Explain the three levels of guidance technology for AGVs.
8. Explain the two approaches to steering AGVs.
9. List and explain three traffic control methods for AGVs.
10. List and explain five methods for dispatching AGVs.
11. List and explain three systems used for monitoring AGVs.

These case studies were provided by the Society of Manufacturing Engineers (SME). They are excerpted from the SME's *Manufacturing Insights*® series of videotapes. These case studies give students actual examples of the way the advanced manufacturing technologies covered in this chapter are being applied in "real-world" manufacturing settings.

As you read each case study, relate the examples cited to the material presented in the text of the chapter. This combination of textbook information and real-world examples will be particularly valuable to your understanding of advanced manufacturing technologies.

GENERAL MOTORS CORPORATION

In Lansing, Michigan, General Motors Corporation uses an automatic guided vehicle (AGV) system to carry parts through the assembly of front-wheel-drive cars. One of the largest AGV systems, it consists of 158 vehicles. It transports engines through assembly; engines and chassis assemblies are merged with car bodies that are moved by an overhead conveyor system.

By using the AGV system, General Motors tests improvements in product quality. The AGV system provides "in-station" assembly, each engine remaining stationary at a particular workstation. Because the AGVs have programmable height adjustment capabilities, the height of the workpiece can be adjusted at each assembly station to fit the requirements of the worker or assembly automation. Improved ergonomics have translated into increased car assembly quality.

The engine dress system is located in the central plant and produces engines for the assembly plants at the rate of 140 engines per hour. The engine dress system contains 95 AGVs and nearly 8,000 feet of guide path. A total of 40 individual workstations are grouped into nine work islands. Only fasteners, tools, and certain small parts are stored at the workstations. All other parts are kitted prior to assembly, placed in tote containers, and manually loaded onto the AGV.

All primary dress assembly operations are done with the engine placed on the AGV. Only minor assembly, such as adding hoses and fastening steps, are done on the traditional monorail conveyor. The operator uses an overhead crane to load each engine onto an AGV. The AGV then moves to the first assembly station, where an assembler uses another

(continued)

overhead crane to add a transaxle to the engine. The AGV then carries the engine through the remaining assembly operations.

One chassis system contains 32 AGVs, the other 33. Each system has approximately 1,000 feet of guide path. The process begins when an operator loads the rear axle and front steering assembly onto an AGV. A specially designed carrier holds the assembly on each AGV.

Each AGV then moves to a workstation where an automated loading device places a one-piece exhaust assembly onto the AGV. Brake lines are installed, then all AGVs move to one of three queue lines where brake systems are filled with fluid. After moving through the brake-testing area, the AGVs wait for computer prompting to meet an overhead conveyor carrying the car bodies. The computer control system releases the matching chassis as the appropriate car body arrives on the overhead monorail.

The AGVs used in the chassis and engine stuff systems are identical, except that the fixtures on the AGVs differ slightly. Automobile engines are mated to bodies as the bodies move on an overhead monorail conveyor. The use of materials handling automation such as this AGV system has brought increased productivity and improved product quality. Other General Motors plants are looking at similar AGV systems to augment their manufacturing capabilities.

3M COMPANY—MAGNETIC DIVISION

The Magnetic Division of 3M Company in Hutchinson, Minnesota, uses a system of truck-mounted conveyors, automated warehousing, bar coding, and automatic guided vehicles (AGVs) to bring higher productivity and quality to the manufacture of videocassettes. The automated materials handling system starts at other plants that produce the molded plastic components of videocassettes.

The videocassette components are nested into specially designed plastic trays using fixed automation. These trays nest one on top of another in stacks of 20 to 30 trays.

Bar codes on the trays contain materials handling sequencing and load information. As the trucks are loaded, this bar-coded information is scanned and downloaded using a phone modem to a computer at the Hutchinson plant. About 80 stacks of trays are loaded on a truck.

Image selected from SME's Automated Material Handling video from the Manufacturing Insights® Video Series.

These trays are moved onto and in the truck using a series of conveyors designed by 3M. Each truck trailer has four lanes of conveyors, two levels in two rows, each lane holding 20 stacks of trays. The conveyors are directed by programmable logic controllers.

The stacks of trays are removed by a fork-lift truck and placed on two input belt conveyor lanes. A bar code reader identifies the containers and part numbers. The bar code also determines the required orientation of the trays for the assembly automation. Additionally, weight and dimensional sizing is automatically determined to ensure that proper quantities have been packaged and the parts are poorly orientated.

These functions are all done automatically under the control of Hewlett-Packard A700 computers. The A700 data is then further integrated into a larger Hewlett-Packard 3000 mainframe computer. The conveyors, sizing, and routing are monitored by Modicon 584 programmable logic controllers.

(continued)

Case Study continued

(continued from page 75)

Components and the parts delivered from other 3M facilities are placed into a two-aisle storage area using automated storage and retrieval equipment. The product manager decides what product to run and inputs this information into the computer system. The computer gives an inventory breakdown of the necessary components in storage available for manufacturing that product.

While production is running, the system monitors all materials handling automation as well as all key assembly stations. This provides for interactive control of the entire manufacturing operation. The conveyor controller directs material to an AGV pickup point. The trays of parts are moved through an airlock onto a waiting AGV. One of four AGVs, each capable of holding 20 stacks of trays, stops at this pickup point. The AGV moves in a guided loop, through a corridor, and into the assembly room.

These last areas are designated clean rooms with special isolation procedures and environmental control systems to insure against airborne and equipment-borne contamination of the product during assembly. In this clean area, the AGVs move to one of three fixed-automation, load/unload stations. Here, stacks of empty trays are loaded onto the AGVs and new parts are unloaded by a conveyor system on the AGV. Other conveyors feed the assembly equipment. The empty trays are then returned to an area for queuing on a conveyor that loads a truck for the return trip to the part molding plants.

3M is a company where the best of the future in materials handling works today—in the use of automation, in its computer integration; a company strong with internal engineering talent that has made its manufacturing solutions work.

A hard copy of the command positions and errors not only provides data to correct or compensate for inaccuracies, but also provides a permanent record of the machine's performance, its limitations, and its sweet spots. Laser alignment and calibration systems also offer definite advantages in accuracy and speed. The real paybacks—scrap reduction, less downtime, and increased production time—help facilities keep costs in line and measure up to the demands of a competitive marketplace.

Major Topics Covered

- CAD/CAM Defined
- Rationale for CAD/CAM
- Historical Development of CAD/CAM
- Computers and Design
- CAD-to-CAM Interface
- CAD/CAM Hardware
- CAD/CAM Software
- Animation and CAD/CAM
- CAD/CAM Selection Criteria
- Group Technology
- *Case Study: K2 Ski Company*
- *Case Study: Flow Systems, Inc.*
- *Case Study: Toro Outdoor Products*
- *Case Study: Oster Division of Sunbeam Corp.*

Chapter Three

◄

Automated Assembly

Automated Guided Vehicles

CAD/CAM ◄

Numerical Control

Industrial Robots

Lasers in Manufacturing

Programmable Logic Controllers

Flexible Manufacturing

Computer Integrated Manufacturing

Other Related Technologies

CAD/CAM DEFINED

This chapter covers the most widely used term in the advanced manufacturing arena. Since the advent of computer technology, manufacturing professionals have wanted to automate the design process and use the database developed therein for automating manufacturing processes. Computer-aided design/computer-aided manufacturing (**CAD/CAM**), when successfully implemented, should remove the "wall" that has traditionally existed between the design and manufacturing components.

CAD/CAM means using computers in the design and manufacturing processes. Since the advent of CAD/CAM, other terms have developed:

- Computer graphics (CG)
- Computer-aided engineering (CAE)
- Computer-aided design and drafting (CADD)
- Computer-aided process planning (CAPP)

These spin-off terms all refer to specific aspects of the CAD/CAM concept. CAD/CAM itself is a broader, more inclusive term. It is at the heart of automated and integrated manufacturing.

A key goal of CAD/CAM is to produce data that can be used in manufacturing a product while developing the **database** for the design of that product. When successfully implemented, CAD/CAM involves the sharing of a common database between the design and manufacturing components of a company.

Interactive computer graphics (ICG) play an important role in CAD/CAM. Through the use of ICG, designers develop a graphic image of the product being designed while storing the data that electronically make up the graphic image. The graphic image can be presented in a two-dimensional (2-D) three-dimensional (3-D), or solids format. ICG images are constructed using such basic geometric characters as points, lines, circles, and curves. Once created, these images can be easily edited and manipulated in a variety of ways including enlargements, reductions, rotations, and movements.

An ICG system has three main components (Figure 3–1): (1) **hardware**, which consists of the computer and various peripheral devices; (2) **software**, which consists of the computer programs and technical manuals for the system; and (3) the human designer, the most important of the three components.

A typical hardware configuration for an ICG system includes a computer; a display terminal; a disk drive unit for floppy diskettes, a hard disk, or both; and input/output devices such as a keyboard, plotter, and printer. Figure 3–2 is an example of an ICG configuration. These devices,

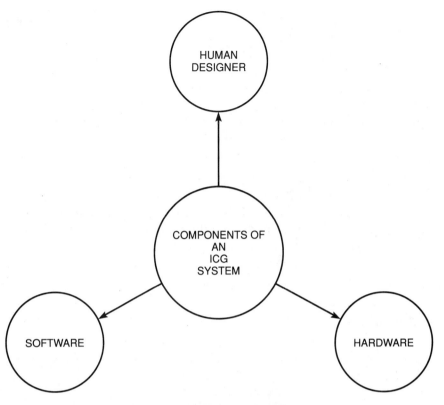

FIGURE 3-1 Components of an ICG system.

FIGURE 3-2 Typical ICG configuration. *Courtesy of CADKEY, Micro Control Systems, Inc.*

along with the software, are the tools modern designers use to develop and document their designs.

Systems such as the one shown in Figure 3–2 enhance the design process by allowing the human designer to focus on the intellectual aspects of the design process, such as conceptualization and making judgment-based decisions. The computer performs tasks for which it is better suited, such as mathematical calculations, storage and retrieval of data, and various repetitive operations such as crosshatching.

RATIONALE FOR CAD/CAM

The rationale for CAD/CAM is similar to that used to justify any technology-based improvement in manufacturing. It grows out of a need to continually improve productivity quality, and, in turn, competitiveness. There are also other reasons why a company might make a conversion from manual processes to CAD/CAM:

1. increased productivity
2. better quality
3. better communication
4. common database with manufacturing
5. reduced prototype construction costs
6. faster response to customers

Increased Productivity

Productivity in the design process is increased by CAD/CAM. Time-consuming tasks such as mathematical calculations, data storage and retrieval, and design visualization are handled by the computer, which gives the designer more time to spend on conceptualizing and completing the design. In addition, the amount of time required to document a design can be reduced significantly with CAD/CAM. All of this taken together means a shorter design cycle, shorter overall project completion times, and a higher level of productivity.

Better Quality

Because CAD/CAM allows designers to focus more on actual design problems and less on time-consuming, nonproductive tasks, product quality improves with CAD/CAM. CAD/CAM allows designers to examine a wider range of design alternatives (e.g., product features) and to analyze each alternative more thoroughly before selecting one. In addition, because labor-intensive tasks are performed by the computer, fewer design errors occur. These all lead to better product quality.

Better Communication

Design documents such as drawings, parts lists, bills of material, and specifications are tools used to communicate the design to those who will manufacture it. The more uniform, standardized, and accurate these tools are, the better the communication will be. Because CAD/CAM leads to more uniform, standardized, and accurate documentation, it improves communication.

Common Database

This is one of the most important benefits of CAD/CAM. With CAD/CAM, the data generated during the design of a product can be used in producing the product. This sharing of a common database helps to eliminate the age-old "wall" separating the design and manufacturing functions (Figure 3–3).

Reduced Prototype Costs

With **manual design**, models and prototypes of a design must be made and tested, adding to the cost of the finished product. With CAD/CAM, 3-D computer models can reduce and, in some cases, eliminate the need for building expensive prototypes. Such CAD/CAM capabilities as solids modeling allow designers to substitute computer models for prototypes in many cases.

Faster Response To Customers

Response time is critical in manufacturing. How long does it take to fill a customer's order? The shorter the time, the better it is. A fast response time is one of the keys to being more competitive in an increasingly competitive marketplace. Today, the manufacturer with the fastest response time is as likely to win a contract as the one with the lowest bid. By shortening the overall design cycle and improving communication between the design and manufacturing components, CAD/CAM can improve a company's response time.

HISTORICAL DEVELOPMENT OF CAD/CAM

The historical development of CAD/CAM has followed close behind the development of computer technology and has parallelled the development of ICG technology. The significant developments leading to CAD/CAM began in the late 1950s and early 1960s. The first of these

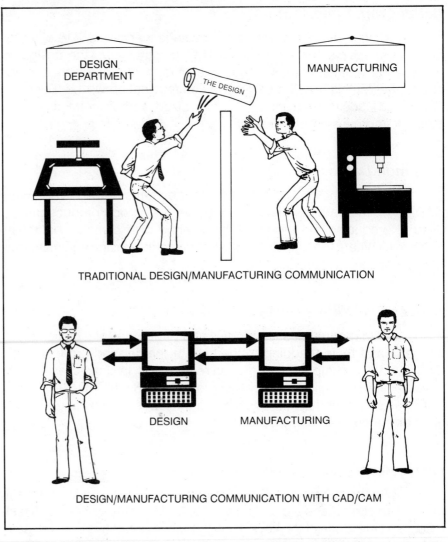

FIGURE 3-3 Traditional design/manufacturing communication compared with CAD/CAM-based communication.

was the development, at Massachusetts Institute of Technology (MIT), of the **Automatically Programmed Tools (APT)** computer programming language.

The purpose of APT was to simplify the development of parts programs for numerical control machines. It was the first computer language to be used for this purpose. The APT language represented a major step toward automation of manufacturing processes.

Another significant development in the history of CAD/CAM followed close behind APT. Also developed at MIT, it was called the

Sketchpad project. With this project, Ivan Sutherland gave birth to the concept of ICG. The Sketchpad project was the first time a computer was used to create and manipulate graphic images on a CRT display in real time.

Throughout the remainder of the 1960s and 70s, CAD continued to develop and several vendors made names for themselves by producing and marketing **turnkey** CAD systems. These were complete systems including hardware, software, maintenance, and training sold as a package. These early systems were configured around mainframe and minicomputers. As a result, they were too expensive to achieve widescale acceptance by small to medium manufacturing firms.

By the late 1970s, it became clear that the microcomputer would eventually play a role in the further development of CAD/CAM. However, early microcomputers did not have the processing power, memory, or graphic capabilities needed for ICG. Consequently, early attempts to configure CAD/CAM systems around a microcomputer failed.

In 1983 IBM introduced the IBM-PC, the first microcomputer to have the processing power, memory, and graphic capabilities to be used in CAD/CAM. This led to a rapid increase in the number of CAD/CAM vendors. By 1989, the number of CAD/CAM installations based on microcomputers equaled the number based on mainframe and minicomputers.

COMPUTERS AND DESIGN

The computer has had a major impact on the way everyday tasks associated with design are accomplished. It can be used in many ways to do many things. However, all design tasks accomplished using a computer fall into one of four broad categories (Figure 3–4):

1. design modeling
2. design analysis
3. design review
4. design documentation

Design Modeling

In CAD/CAM **design modeling**, a **geometric model** of a product is developed that describes the part mathematically. This mathematical description is converted to graphic form and displayed on a **cathode ray tube (CRT)**. The geometric model also allows the **graphic image** to be easily edited and manipulated once displayed.

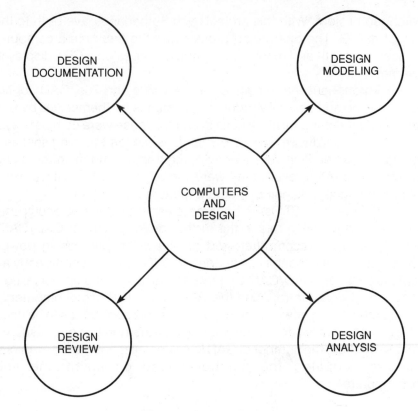

FIGURE 3–4 Computers and design.

The process begins when a designer creates a graphic image using a special ICG software package. The image is created by entering commands that cause the computer to construct the image out of points, lines, circles, and curves. To create a graphic image, the computer must translate the geometric characters into a corresponding **mathematical model**. What appears to the designer as a graphic image is stored in the computer as a series of mathematical coordinates.

As the designer issues commands to edit or manipulate a graphic image, the computer revises the mathematical model. The computer must first change the geometric model before it can change the graphic image, and it must calculate the mathematical coordinates of a geometric character before it can display that character.

Graphic images are displayed in one of three formats:

1. two-dimensional or 2-D
2. two-and-a-half-dimensional or 2½-D
3. three-dimensional or 3-D

These three formats are illustrated in Figures 3–5, 3–6, and 3–7. A 2-D graphic image shows an **orthographic** (flat) **representation** and usually shows two or more views (e.g., top, front, and right side). A 2½-D graphic image is an **oblique representation**. A 3-D graphic image may be a wireframe model or a true 3-D solid model (Figure 3–8).

Communication can be enhanced further when graphic images are displayed in color. Many capabilities of CAD/CAM, including finite-element analysis, wireframe 3-D, and solids modeling, are more easily understood when the graphic images that result are displayed in color. Specific applications such as piping drawings are also easier to read when displayed in an appropriate variety of colors.

When graphic images are displayed in only one color, it is sometimes difficult to determine when lines cross over or under, what is near and what is far away, what is the original image and what is the distorted image (as in the case of finite-element analysis), and what layer a given component is on. The use of color allows for a more definitive distinction between over and under, near and far, and original and distorted. It also

Mechanical Drawing

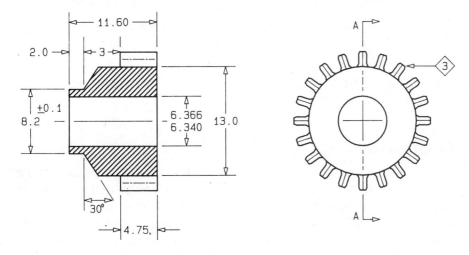

Section View   Top View

Plotted on the Hewlett-Packard 7550 Graphics Plotter using the IBM PC

FIGURE 3–5 2-D drawing. *Courtesy of VERSACAD Corporation.*

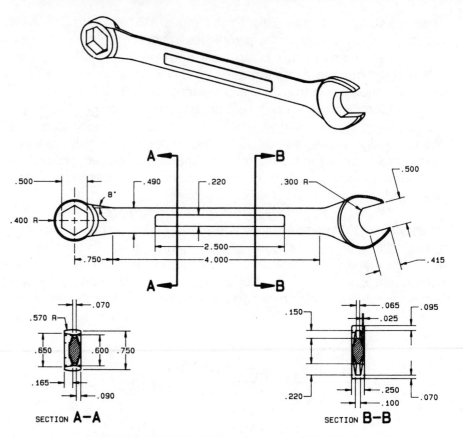

FIGURE 3–6 2½-D drawing. *Courtesy of Tektronix, Inc.*

helps clarify objects consisting of numerous surfaces or several component parts.

Design Analysis

The computer has simplified the **design analysis** stage of the design process significantly. Once a proposed design has been developed, it is necessary to analyze how it will stand up to the conditions to which it will be subjected. Such analysis methods as heat transfer and stress-strain calculations are time consuming and complex. With CAD/CAM, special computer programs written specifically for analysis purposes are available.

One such program that has simplified the analysis of manufactured products is called **finite-element analysis**. Finite-element analysis

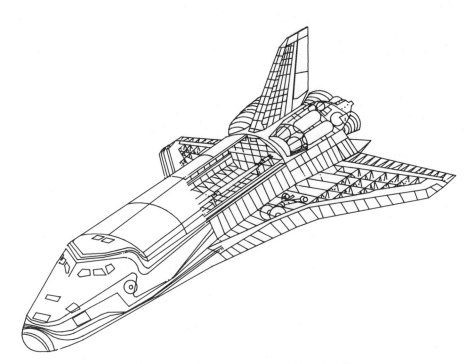

FIGURE 3–7 3-D drawing. *Courtesy of Autodesk, Inc.*

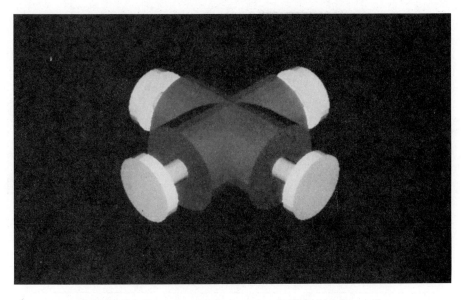

FIGURE 3–8 3-D solid model. *Courtesy of VERSACAD Corporation.*

involves breaking an object up into many small rectangular or triangular elements (Figure 3–9), then analyzing each individual element by computer. This approach gives a thorough analysis that might not be feasible without the aid of a computer. It also offers the advantage of specifically pinpointing the locations of problems so that design corrections can be more easily made.

With finite-element analysis, the computer must have a powerful processing capability. The computer analyzes the whole object by analyzing each individual interconnected finite element. By analyzing the response of each finite element of the object to the stress, strain, heat, or other force acting on it, the computer can predict the reaction of the entire object.

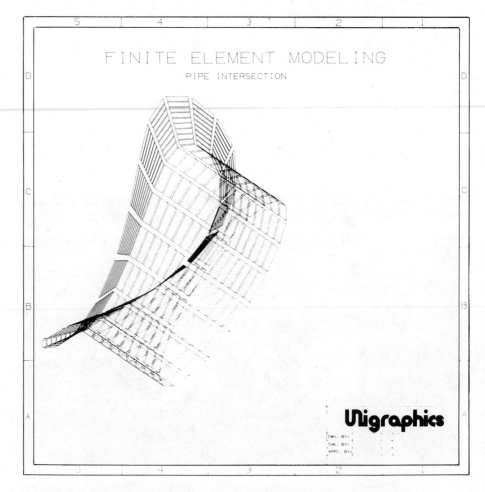

FIGURE 3–9 Finite-element model. *Courtesy of McDonnell Douglas Information Systems Company.*

Modern CAD/CAM systems with finite-element analysis capability make the process simple to achieve. Users define the area that is to be divided. The computer then automatically divides the area into the interconnected network of finite elements. A particularly valuable characteristic of finite-element analysis is its ability to visually display the results of the analysis.

For example, if a part is to be analyzed to determine how it will behave when subjected to a specified amount of stress, the computer is able to superimpose the image of the stressed part over the unstressed part. In this way, the resultant distortion can be easily seen. Such visual evidence of the results can make it easier for the designer to pinpoint the necessary design changes.

Design Review

Another step in the design process that has been simplified by the computer is **design review**. This involves checking the accuracy of all aspects of the design. There are several ICG capabilities that make design review in CAD/CAM easier than with manual design.

The first is the **semiautomatic dimensioning** capability of many CAD/CAM software packages. To produce a graphic image, an ICG system must first create a mathematical model. The data in this model can be used by the computer for calculating dimensions automatically. Most good CAD/CAM software packages can display these dimensions according to standard dimensioning and tolerancing practices. This means fewer dimensioning errors, a common problem with manual dimensioning.

The **layering** capability of CAD/CAM software also simplifies design review. If a multisided printed circuit board has been designed, the traces for each successive layer can be displayed separately in a different color on a CRT display. They can also be displayed simultaneously. In this way, traces can be easily checked to ensure that all connections have been properly layed out. The layering capability can also be used to check complex piping layouts to ensure that pipes do not intersect where they should not and do intersect where they should. Good CAD/CAM systems can operate on over 250 layers.

Another CAD/CAM software capability that has simplified design review is called **interference checking**. With this capability, mating parts can be joined on the CRT display as they would be in a final assembly. The designer can see immediately if there is interference. With interference checks, the computers **zoom** capability is particularly useful.

Zoom allows the designer's eye to move progressively closer to small, intricate details in a drawing. It gives the appearance of magnifying small details so they can be seen more easily. This is particularly useful when working on complex assemblies made up of numerous small subassemblies.

Some of the more advanced CAD/CAM software packages have a **kinematic** capability. This means they can simulate motion on a CRT display. This is particularly useful when working on a design made up of moving parts. Computer simulation in such cases takes the place of the cardboard and plastic models used by designers in the past.

Design Documentation

Design documentation is another area in which CAD/CAM offers major benefits when compared with manual drafting techniques. With CAD/CAM, the drawings needed to document the design can be produced using the database created during the design process. Five-fold improvements in drafting productivity are now common with modern CAD/CAM systems.

CAD/CAM software has simplified such time-consuming drafting tasks as dimensioning, creating magnified views of intricate details, **transformations** (e.g., isometric, dimetric, and trimetric projections), scaling, correcting errors, and making design revisions. It also made the concept of **nonrepetition** a reality in drafting.

One of the most unproductive tasks associated with manual drafting is redrawing details, sectional views, or other drawings that have been previously drawn. Because the computer can store the mathematical models of all drawings done on a CAD/CAM system, once a drawing has been produced, it never has to be redrawn. It can simply be called up from storage, entered into the appropriate location in the drawing package, and used again and again.

CAD-TO-CAM INTERFACE

With CAD/CAM, the real interface between the design and manufacturing components is the common database they share. This is the essence of CAD/CAM. With manual design and manufacturing, engineers go through each step in the design process, drafters produce drawings and other documents to communicate the design, manufacturing personnel use the drawings to develop process plans, and shop personnel actually make the product.

With the old approach, until the design and drafting personnel completed their work, the manufacturing personnel did not see it. The design and drafting department did its job and "threw the plans over the wall" to manufacturing so it could do its job. This approach led to continual breakdowns in communication as well as poor relations between the design and manufacturing components. The result was a loss of productivity.

With CAD/CAM, manufacturing personnel have access to the data created during the design phase as soon as they are created. At any point in the design process, they can call up information from the design database and use it. Since the data are shared from start to finish, there are no surprises when the completed design is ready to be produced. While designers are creating the database and drafters are documenting the design, manufacturing personnel can be planning, setting up, ordering materials, and preparing parts programs.

Everything needed by manufacturing personnel to produce the product is contained in the common database. The mathematical models, graphic images, bills of material, parts lists, size, form, locational dimensions, tolerance specifications, and material specifications are all contained in the common database (Figure 3–10).

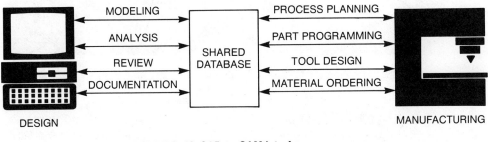

FIGURE 3–10 CAD-to-CAM interface.

CAD/CAM HARDWARE

The hardware part of a CAD/CAM system is typically configured to include a **central processing unit (CPU)**, various output devices such as plotters and printers, a secondary storage capability, a graphics terminal, and various input devices (Figure 3–11).

The CPU for a CAD/CAM system is the same as a CPU for any computer-controlled system. It contains the main memory and logic/arithmetic section for the system. The most widely used **secondary storage** medium in CAD/CAM is the hard disk, floppy diskette, or a combination of both.

Designers actually work at a CAD **workstation**. A workstation consists of the hardware components needed to produce and document

FIGURE 3–11 Typical CAD/CAM configuration. *Courtesy of VERSACAD Corporation.*

a design. A typical CAD workstation (Figure 3–12) consists of the graphics terminal and input devices needed to interact with the system. Plotters and printers may be nearby or in remote locations where they are shared by several workstations.

Graphics Terminals

There are a number of different types of graphics terminals that can be used in a CAD/CAM system. The three most widely used types are:

1. directed-beam refresh
2. direct-view storage tube
3. raster scan

Directed-beam refresh terminals create a graphic image by directing an electron beam against a phosphor-coated CRT display (Figure 3–13). Wherever the electron beam comes in contact with the phos-

FIGURE 3–12 Typical CAD/CAM workstation. *Courtesy of VERSACAD Corporation.*

phor coating, the coating illuminates. By focusing the electron beam in such a way as to control the various points of contact, the desired graphic image can be created. This method of creating a graphic image is also called "stroke writing."

Once the directed beam creates an image, the illuminated areas must be continually redrawn or refreshed. If this does not happen, the image will fade and flicker. **Refreshing the image** must be done so frequently that the cycle is measured in microseconds. An advantage of directed-beam refresh technology is that graphic images can be easily revised.

Direct-view storage tube (DVST) terminals are similar to directed-beam refresh terminals with one key exception; refreshing the image is not necessary. As the name implies, once the electron beam has illuminated an image, the screen is able to "store" or retain the image. The image is not actually stored; it is maintained by an electron flood gun that floods the illuminated area, causing the image to hold its density. A disadvantage of DVST technology is that graphic images cannot be revised as readily as they can with refresh technology.

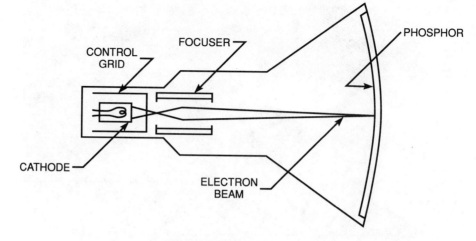

FIGURE 3–13 Cathode ray tube.

Raster scan terminals create images by illuminating tiny picture elements or **pixels** on the screen. With this technology, the screen is divided into columns and rows of pixels. Each pixel can be illuminated by an electron beam. The greater the number of pixels, the higher is the quality of the graphic image. This concept is known as **resolution**. High-resolution terminals can produce high-quality, smooth, accurate images. Low-resolution terminals produce rough, low-quality images.

With a raster scan terminal, the actual image is created when an electron beam scans the columns and rows of pixels horizontally from left to right, illuminating the appropriate pixels as it goes (Figure 3–14). Color terminals can illuminate pixels in different colors and different degrees of brightness. The typical raster terminal can make over 50 scans per second.

Raster scan technology is similar to television technology. The basic difference is that television images are created by analog signals generated by a video camera, while raster scan terminals create images based on digital signals generated by a computer.

Color terminals are popular in modern CAD/CAM systems. A color capability can significantly enhance communication in the CAD component. Color images are created using three electron beams; one for each of the three basic colors of red, green, and blue. By combining these three colors in differing intensities, some terminals are capable of producing a wide variety of colors. Figure 3–15 shows the colors typically available with modern color terminals. Raster scan terminals produce the best color images of the three main types of terminals because they are able to vary the color capabilities of each pixel and because of the mixing of the three electron beams.

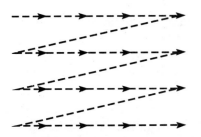

FIGURE 3-14 Portion of a scan cycle.

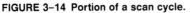

RED	BLUE
YELLOW	MAGENTA
GREEN	WHITE
CYAN	

FIGURE 3-15 Colors typically used in CAD/CAM.

Input Devices

      Designers interact with CAD/CAM terminals using a wide variety of input devices. These various devices fall into three broad categories (Figure 3-16):

1. keyboards
2. digitizers
3. cursor control devices

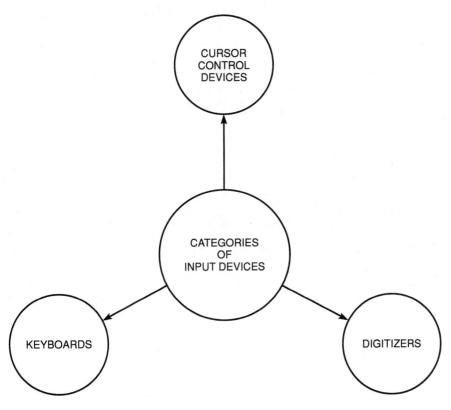

FIGURE 3-16 Categories of input devices.

Keyboards are used for entering alphanumeric data and issuing commands. Commands are usually arranged within menus that can be displayed on the terminal screen or on an electronic pad. Figure 3–17 shows examples of the types of menus used in modern CAD/CAM systems.

Cursor control devices are used to create, position, identify, and manipulate graphic data. Over the years, a number of different cursor control devices have been developed for use in CAD/CAM systems. These include joysticks, thumbwheels, trackballs, direction keys, light pens, and electronic tablets.

Joysticks and trackballs (Figure 3–18) are no longer widely used. Joysticks are the same type of device used with many video games. The stick can be pushed through an angle causing the cursor to

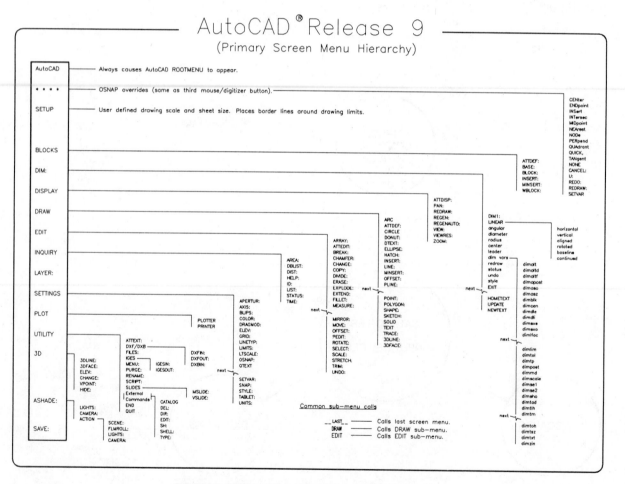

FIGURE 3–17 Menu hierarchy. *Courtesy of Autodesk, Inc.*

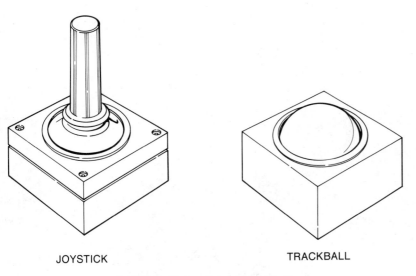

JOYSTICK   TRACKBALL

FIGURE 3-18 Joystick and trackball.

move correspondingly. Trackballs are also used with video games. The ball is turned as desired and the cursor moves correspondingly.

Thumbwheels and direction keys are contained either on the keyboard or the terminal console. Both are used to move the cursor horizontally and vertically. These devices are still used, but not widely.

Two widely used cursor control devices are the **light pen** and **electronic tablet**. The term light pen is deceptive because the device does not emit light; it detects light on the terminal screen using a sensor contained in the device. When a light pen touches a terminal screen, light is sensed at a given location and the computer calculates the position on the screen. It then takes whatever action is associated with that position. Figure 3-19 shows a light pen being used.

Electronic tablets such as the one shown in Figure 3-20 are used along with a stylus or puck for positioning the cursor. The flat surface corresponds with the terminal screen. By moving the stylus to different locations on the tablet, a designer moves the cursor correspondingly.

In addition to cursor control, light pens and electronic tablets can be used to select options from screen or tablet mounted menus, move graphic images on the screen, and select portions of the screen for various data manipulations.

Digitizers are large electronic boards that look like traditional drafting boards (Figure 3-21). They are used to convert graphic data into digital data. This process allows drawings produced manually or through

FIGURE 3–19 Light pen use. *Courtesy of CalComp Inc.*

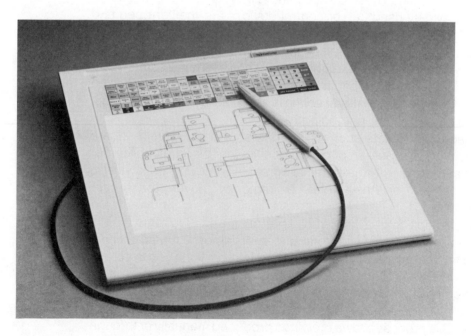

FIGURE 3–20 Electronic tablet. *Courtesy of Calcomp Inc.*

FIGURE 3–21 Digitizers. *Courtesy of CalComp Inc.*

some other method to be entered into a CAD/CAM system. Figure 3–22 shows the required and optional hardware for a typical modern CAD system.

CAD/CAM SOFTWARE

Software allows the human user to turn a hardware configuration into a powerful design and manufacturing system. Computer graphics software allows users to accomplish tasks in five broad categories of functions (Figure 3–23):

1. producing graphic images
2. manipulation of graphic images
3. controlling the terminal display
4. performing editing functions
5. performing input functions

Required Hardware

A minimally configured CADKEY workstation requires the following hardware: IBM PC, PC/XT, PC/AT microcomputer (or compatible) with MS DOS 2.0 or higher, 512k memory, and a hard disk with one floppy drive. (A math co-processor, 640k, and 2 serial ports are strongly recommended.)

Color Graphics Adapter Card
(to operate a graphics monitor) CADKEY supports:

- AT&T Display Enhancement Board—16 colors (640 x 400)
- AT&T Indigenous Graphics Board—black & white (640 x 400)
- Conographics Conocolor 40—16 colors (640 x 400)
- Hercules Graphics Card—monochrome (720 x 348)
- IBM Color Graphics Adapter—black & white (640 x 200)
- IBM Enhanced Graphics Adapter—4 or 16 colors (640 x 350)
- Imagraph-AGC4 Series—16 colors (1280 x 1024), (1024 x 768)
- Metheus Omega/PC—16 colors (1024 x 768)
- Number Nine-Revolution 2048 x 4 Series 16 colors—(1280 x 960), (1024 x 768), (832 x 624)
- Sigma Color Designs 400—16 colors (640 x 400)
- Tecmar Graphics Master—16 colors (640 x 400)
- Tecmar Graphics Master—monochrome (720 x 704)

Require 640k

- IBM Professional Graphics Adapter—16 colors (640 x 480)
- Verticom M-16—16 colors (640 x 480)

Plus other cards which emulate one of the above. *Call us for additional cards not listed.*

Optional Hardware

Digital Pen Plotters, CADKEY supports:

- Calcomp 1040 series (PCI, 907, 951 controllers)
- Gould 6120 (HPGL)
- Hewlett Packard HP7400 and 7500 series
- Houston Instruments, DMP series
- IBM 7372 (HPGL)
- Ioline LP 3700 (DMPL)
- Nicolet/Zeta (HPGL)
- Numonics (HPGL)
- Roland DXY-880 (HPGL)
- Sweet-P 600 (HPGL)
- Western Graphtec MP2000 (HPGL)

Printer, CADKEY supports:

- Diablo C-150, C-200
- Epson MX, RX, FX, JX-80
- Fujitsu DL 2600C
- Hermes PC Print 4
- HP Laser Jet 2686A
- MPI Print Mate 350
- Toshiba P-351C

2-D Digitizers and Macro Tablets, CADKEY supports:

- Calcomp 2000, 2500
- Cherry Tablet
- GTCO Digipad 5 Lo/Hi Res Tablets
- Hitachi Tiger
- Houston Instrument HI-PAD DT-11
- Houston Instruments True Grid
- Kurta Series 1 and 2
- Numonics 2200 series
- Polytel Keyport Macro Tablet 60,300
- Scriptel SPD series
- Summagraphics Bit Pad One, Two, Model 1103
- Summagraphics MM Series Models 961, 1201, 1812
- Summagraphics Microgrid
- Pencept Penpad

Mouse, CADKEY supports:

- Logitech Mouse
- Microsoft Mouse
- Mouse Systems Mouse
- Summagraphics Summamouse
- Torrington Mouse

CADKEY,® Division of MICRO CONTROL SYSTEMS, INC.
27 Hartford Turnpike, Vernon, Conn. 06066 • Tel. (203) 647-0220 • TWX: 5106007223

FIGURE 3–22 Hardware specifications for a CAD system. *Courtesy of CADKEY, Micro Control Systems, Inc.*

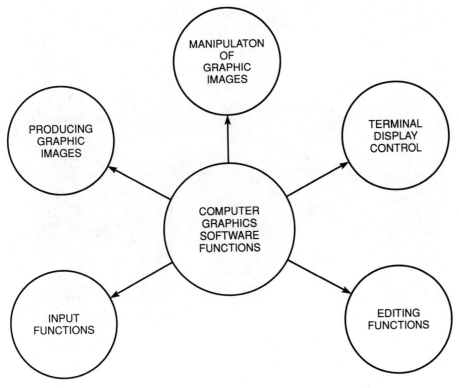

FIGURE 3–23 Computer graphics software functions.

Producing Graphic Images

All design documentation, whether in the form of drawings, bills of material, parts lists, or specifications, consists of geometric characters and alphanumeric data. Alphanumeric data are data comprising letters, numbers, and other special keyboard figures. Geometric characters include points, lines, circles, and the other components graphic images.

All of these elements are incorporated into computer graphics software in a way that allows designers to use them as building blocks in producing documentation packages. Figure 3–24 contains examples of simple graphic images made of basic geometric characters.

Manipulating Graphic Images

Once a graphic image has been produced, designers must be able to manipulate it. Manipulating a graphic image means enlarging, re-

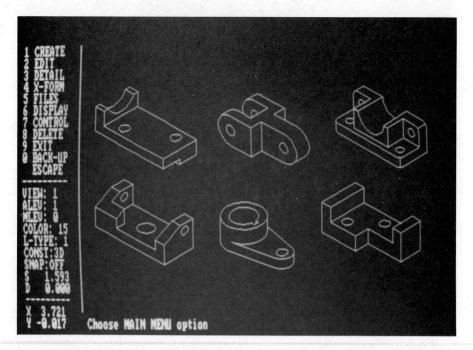

FIGURE 3–24 Simple line drawing. *Courtesy of CADKEY, Micro Control Systems, Inc.*

ducing, rotating, or moving the image or portions of it. Such manipulations are accomplished by transformations in which the mathematical model of the image is altered or repositioned in the database.

Controlling The Terminal Display

One of the advantages of ICG is a capability known as **windowing**. Windowing is a display control concept that allows users to place a window in any location on the display so that data therein can be controlled and manipulated. Data within a specified window might be rotated, enlarged, or reduced.

Another display control function is **hidden line removal**. Anytime a graphic image is displayed in a wireframe 3-D format, some of the lines are visible and some are hidden. When both visible and hidden lines are superimposed on each other, a graphic image can be difficult to read and understand. Through hidden line removal, hidden lines are eliminated from the image, making the image easier to read and understand. Figure 3–25 is an example of a complex 3-D drawing that would be difficult to read if the hidden lines had not been removed.

FIGURE 3–25 3-D drawing. *Courtesy of VERSACAD Corporation.*

Performing Editing Functions

Editing a graphic image means revising or correcting it by deleting, replacing, or otherwise changing part of it. These parts are known as **segments**. Consequently, editing a graphic image is sometimes called segmenting.

Editing in CAD is the same thing as erasing and redrawing in manual documentation. It could mean removing a line and replacing it, adding new lines and geometric characters, changing alphanumeric data, or any of the other tasks associated with revising or correcting a drawing.

Performing Input Functions

The computer graphics software package dictates the way users interact with a system, input data, and issue commands. It also dictates the input devices that can be used. The better computer graphics software packages are typically the easiest to interact with. They are **user friendly.**

One of the easiest ways to issue commands and otherwise interact with a CAD/CAM system is through the use of menus built into the software package (Figure 3–26). Menus simplify input functions by allowing users to simply select the actions they want the system to undertake.

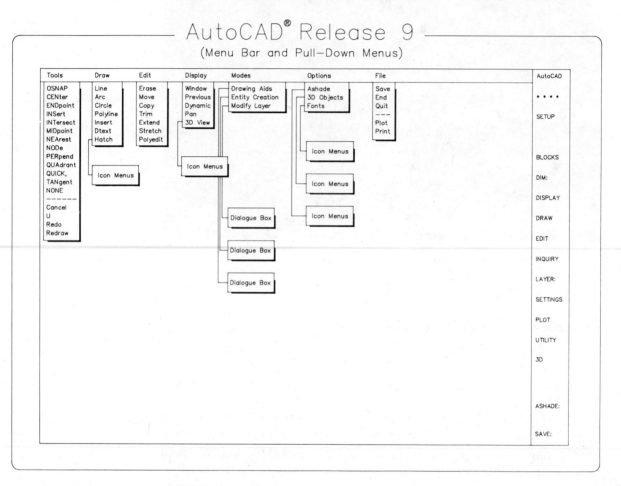

FIGURE 3–26 Pulldown menus. *Courtesy of Autodesk, Inc.*

CAD/CAM Database

Perhaps the most important element of a computer graphics software package is the database. It is the potential for sharing a common database that is the heart of CAD/CAM. The CAD/CAM database contains the following:

ICG SOFTWARE
- System commands
- Menus
- Output routines

DOCUMENTATION CREATED USING THE SOFTWARE
- Drawings
- Design information
- Bills of material
- Parts lists
- Specifications
- Other items of documentation

It is the data in the second category, documentation, that are communicated to manufacturing personnel and used in the production of the design.

2-D Versus 3-D Software

CAD/CAM systems are available with both 2-D and 3-D capabilities. Early systems were limited to 2-D. This was a serious shortcoming because 2-D representations of 3-D objects is inherently confusing. Historically, many manufacturing errors and much waste have been associated with an inability of design personnel to properly represent an object in a 2-D format. Equally problematic has been the inability of manufacturing personnel to properly read and interpret complicated 2-D representations of objects.

Two-dimensional drawings generally show two more orthographic (flat) views (e.g., the top, front, and right side) of the object. Figure 3–27 is an example of a 2-D drawing showing only one view. While this drawing is attractive and accurate, it does not contain enough information to allow manufacturing personnel to produce it without consulting with design personnel, a critical test that all drawings must pass.

At least two additional views, perhaps a top and side, would be needed for manufacturing personnel to have the information they would need to produce it. Even with these additional views, their task would be difficult because the human eye does not see in 2-D nor does the human mind think in 2-D. Therefore, even with the necessary views, manufacturing personnel could find it difficult to visualize the finished product. Accurate visualization skills take intensive training to learn and years of experience to perfect, which limits the productivity of both design and manufacturing personnel in a 2-D documentation setting.

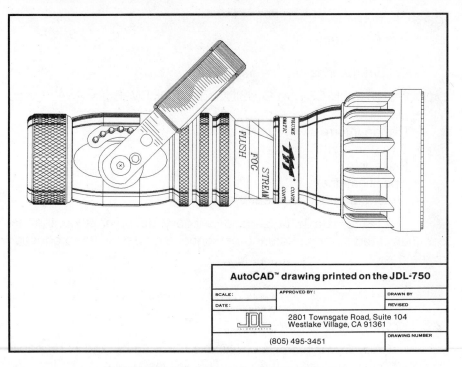

AutoCAD™ drawing printed on the JDL-750

SCALE:	APPROVED BY:		DRAWN BY
DATE:			REVISED
JDL INCORPORATED	2801 Townsgate Road, Suite 104 Westlake Village, CA 91361		
	(805) 495-3451		DRAWING NUMBER

FIGURE 3–27 2-D drawing. *Courtesy of Autodesk, Inc.*

The trend in CAD/CAM is toward 3-D representation of graphic images. Such representations approximate the actual shape and appearance of the object to be produced; therefore, they are easier to read and understand.

Computer graphics software with 3-D capabilities is of two types: wireframe and solid 3-D models. Three-dimensional representations of objects are called models. **Wireframe models** are less sophisticated than solid models and require less computer memory and processing power to support. They approximate the appearance of the real object through **axonometric** (isometric, dimetric, and trimetric) **representations**. Figures 3–28 and 3–29 are examples of wireframe models. They are easier to read and understand than 2-D models, but even these are not true representations of the actual object. The most critical weakness of wireframe models is in the representation of contoured surfaces and edges. It can be difficult to determine whether a surface is flat or contoured, solid or open.

This weakness led to the development of solids modeling as a CAD/CAM capability. A **solid model** is a computer model of the actual

FIGURE 3–28 3-D wireframe model. *Courtesy of VERSACAD Corporation.*

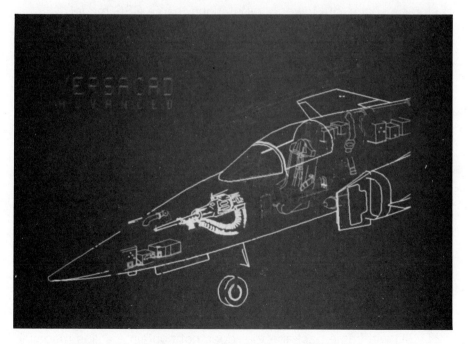

FIGURE 3–29 3-D wireframe model. *Courtesy of VERSACAD Corporation.*

object, inside and out, built in the database and displayed on the terminal screen. Figures 3–30 and 3–31 are examples of 3-D solid models.

There are two basic approaches to solid modeling:

1. constructive solid geometry
2. boundary representation

Constructive solid geometry (CSG), sometimes called the building block approach, uses simple solids such as cubes, pyramids, cylinders, and spheres to construct the model. CSG is considered to be easy to use and understand. It is not unlike building a live model out of wooden building blocks, something many people have done and can relate to. Another positive aspect of CSG is that the mathematical model housed in the database is compact and precise. It requires a great deal of computation, but relatively little storage space.

Boundary representation (B-rep), involves actually drawing the boundary of the model so that it appears on the display of a graphics terminal. Such drawings are orthographic and include all of the views (top, front, right side, bottom, back, and left side) needed to fully define the object. The B-rep approach is more effective than CSG in constructing oddly shaped models that cannot be easily made of standard geometric building blocks. A B-rep model requires a relatively large amount of storage space, but fewer computations than a CSG model.

The most modern solids modeling software packages combine the CSG and B-rep approaches to take advantage of the strengths of each while minimizing the weaknesses. This capability has enhanced communication within the design component and between the design and manufacturing components. But it does come with a price. Solids modeling software is expensive in two ways. First, the software package itself is expensive. Second, it takes much more sophisticated hardware to support it. In spite of this, solids modeling software represents the future of CAD/CAM. There is a strong consensus among design and manufacturing professionals that solids modeling simplifies communication and, as a result, decreases errors and waste. Computer technology continues to develop rapidly enough to ensure that the hardware needed to support solids modeling software will become continually less expensive, as will the software itself. Finally, solids modeling can eliminate some of the need for the construction of expensive live models and prototypes. Figures 3–32 through 3–35 are examples of solid models constructed on a modern CAD/CAM system. Figure 3–36 lists the standard features of a widely used CAD/CAM software package. Note that this is a 2-D/3-D wireframe package that does not contain a solids modeling capability.

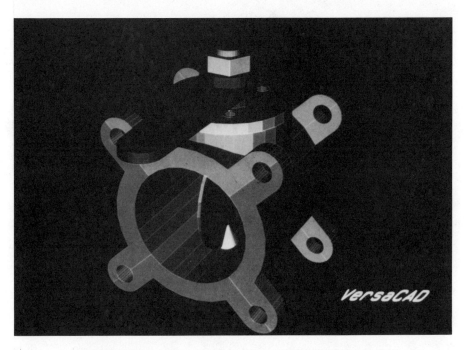

FIGURE 3–30 3-D solid model. *Courtesy of VERSACAD Corporation.*

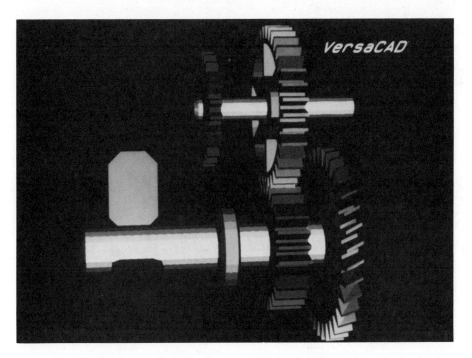

FIGURE 3–31 3-D solid model. *Courtesy of VERSACAD Corporation.*

FIGURE 3–32 3-D solid model. *Courtesy of VERSACAD Corporation.*

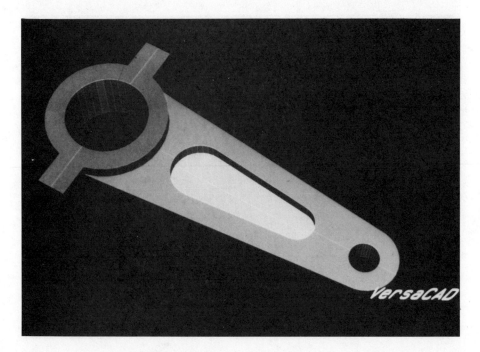

FIGURE 3–33 3-D solid model. *Courtesy of VERSACAD Corporation.*

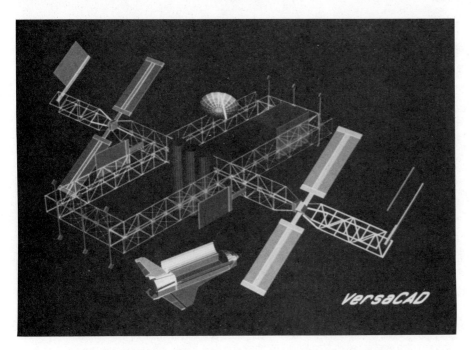

FIGURE 3–34 3-D solid model. *Courtesy of VERSACAD Corporation.*

FIGURE 3–35 3-D solid model. *Courtesy of VERSACAD Corporation.*

Fully integrated 2-D/3-D drafting and design system with a schema-based menu structure. The CADKEY system offers:

- Compliance to ANSI and ISO standards.
- A standard cartesian coordinate system (x, y, z).
- Efficient selection routines with options such as single select, window, chain, groups, all displayed and several types of masking including masking by level.
- A universal Position Menu offering multiple methods for indicating a position (or location) in 3-D space.
- Instant accessibility to desired functions from anywhere within the menu structure via assigned Immediate Mode commands or status window.
- 256 levels
- 16 colors
- High resolution display capability.
- Open database for easy access with external programs.
- Compatible with over 10,000 combinations of available computer hardware.
- Ability to execute a DOS command or another program from within CADKEY environment.

Geometric Modeling

- Basic drawing components include lines, arcs, points, circles, 2-D/3-D cubic parametric splines, line mesh, chamfers, polygons, point mesh, autoseg, rectangles and fillets with thousands of methods of construction. For example, a line may be created using any of the following methods: designating start and endpoints, continuous endpoints, parallel or perpendicular to another line, horizontal and/or vertical direction, at any angle, mesh, sketch mode, or tangent to an arc and point or two arcs, and many more.
- Trimming, extending or breaking of selected line and curve entities.
- XFORM (transform) options for scaling, translation, rotation, extrusion, and mirroring of geometry.
- Continuous tracking of cursor location (x, y position).
- Semi-automatic hidden line removal.

Detailed Dimensioning

- Automatic dimension of the angle between two lines, the diameter and/or radius of a circle or arc, or the horizontal, vertical or parallel dimension between two points.
- Full user control over setting or changing parameters and values for new or existing dimensions, labels or notes.
- Automatic dimension tolerancing which may be set or changed at any time.
- User selectable units for dimensioning (inches, feet, centimeters, meters, etc.).
- User-specified text alignment and location with variable font types.
- Level independent line types, text fonts, line widths, colors and pen numbers.
- Several different crosshatching styles, meeting ANSI standards, are offered with user-defined angle of rotation and scale.
- Read-in disk file notes.

Display Manipulation

- Cursor snapping and grid display at any time with ability to re-align or modify grid or snap position or increment size.
- Unlimited amount of user-defined views as well as 8 pre-defined views already offered by the system: top, front, back, bottom, right, left, isometric, axonometric.
- Precise detailing is permitted with the Zoom and Pan features.

Entity Management

- Entity deletion by indicating entity position, type(s) or level(s).
- Recall of any or all deleted entities and levels.
- More than one geometric or dimension entity may be translated at a given time. Desired entities are surrounded by a "rubberbox" window and translated to a new position via base position or x, y coordinates.
- User-defined patterns and entities may be grouped together to form an integrated unit, callable as a full design group as desired.

Geometric Analysis

- Area, perimeter, moment of inertia and centroid are automatically calculated based on selected entities.

(over)

Entity Verification
- Existing entities may be examined at any time for current attribute, coordinate data, distance, length or angle assignments.

File Management
- Up to five types of files supported: Part, Pattern, Plot, CADL (CADKEY Advanced Design Language) and DXF (Drawing Interchange Format).
- Complete Part and Pattern file creation and retrieval.
- Input/Output compatibility with 2-D drafting files allowing complete 2-D geometry and text exchange from one system to another (DXF).
- CADL, CADKEY's Advanced Design Language file contains input or output data primitives with certain design language constructs allowing parameterized design programs. This provides easy database access.
- Complete file conversion program for version updates.
- Device independent plot file management.

And More...
- IGES translator for bi-directional transfer of 3-dimensional data.
- Off-line plotting—independent of dedicated workstation.
- On-line plotting and printing at User-specified units.
- Complete and updated user documentation and tutorial.
- Customization Guide.
- Training courses are available along with an in-depth training manual.
- CADKEY is marketed in Germany, Switzerland and Austria under the CADSTAR trademark.
- CADKEY is available in English, German, French, Spanish and Italian.

For further information call CADKEY at (203) 647-0220, Telex (TWX) number 5106007223.

FIGURE 3–36 **Standard software features.** *Courtesy of CADKEY, Micro Control Systems, Inc.*

ANIMATION AND CAD/CAM

Kinematic simulation or **animation** can be an important CAD/CAM capability in certain settings. Through animation, users can design a product or a process that involves moving components and analyze its behavior without having to build prototypes and conduct live trial runs.

For example, mechanical linkage designs have historically been difficult because testing their actual behavior has required engineers to build models or prototypes. In such a setting, engineers frequently resort to building cardboard or wooden models, a time-consuming process. With animation, a computer model can be quickly built and displayed. By watching the computer screen, the engineer can easily see how the linkage performs.

Other useful applications of animation are in analyzing the performance of a robot in a cell or automated guided vehicles (AGVs) on the shop floor. By simulating the planned behavior of a robot or an AGV, engineers and manufacturing personnel can spot and correct problems before implementing a live setting.

CAD/CAM SELECTION CRITERIA

Selecting the best hardware and software for a given CAD/CAM setting can be difficult, but it is critical to the success of CAD/CAM. For the inexperienced, the turnkey approach is recommended. This means purchasing a CAD/CAM system that comes with everything necessary for implementation, including hardware, software, installation, and training. The best system will vary, depending on the needs and intended applications of the individual company.

Before contacting vendors, it is a good idea to develop an **applications checklist**. This is a comprehensive list of everything the system is expected to do. For example, if an electronics firm were to develop an applications checklist, it might include the following:

DESIGN DOCUMENTATION
1. schematic diagrams
2. circuit diagrams
3. logic diagrams
4. wiring diagrams
5. integrated circuit diagrams
6. printed circuit board drawings
7. bill of materials preparation
8. design rules checking

SUPPORTIVE DOCUMENTATION
1. specifications
2. technical manuals
3. charts and graphs
4. flow charts

These are only a few of the items that might be included on the applications checklist of an electronics manufacturing firm. A complete checklist would be much more comprehensive. In putting together such a checklist, it is important to convene a team of experienced personnel who are thoroughly knowledgeable of the company's needs.

Once the applications checklist is complete, vendors may be contacted for bids or quotes. The first step, of course, is to rule out systems that cannot handle everything on the checklist. Once the list of vendors is narrowed down to those with appropriate systems, the next step can be taken. In this step, criteria in the following categories are used to select the system to be purchased:

1. initial costs
2. ongoing costs
3. services and support
4. quality
5. delivery and installation

Initial Costs

In this category, the initial costs of hardware and software are examined and compared, including the costs of the following:

- CPU
- Workstations (initial)
- Workstations (added later)
- Peripherals (initial)
- Peripherals (added later)
- Supplies
- Facility renovations
- Basic software package
- Additional software packages
- Special documentation

Ongoing Costs

In this category, the ongoing costs associated with the system are examined and compared, including the costs of the following:

- Hardware maintenance
- Software maintenance
- Field support
- Licensing fees
- Supplies
- Spares

Services and Support

In this category, the various services that might be provided by a vendor are examined and compared. This is critical. Buying a CAD/CAM system is similar to buying an automobile in that the buyer will need ongoing services from the dealer. CAD/CAM vendors who sell systems but do not provide ongoing services should be eliminated from consideration. Services to look for include the following:

- Hardware maintenance
- Software maintenance
- Prompt field support
- Debugging of software
- Training (initial and advanced)
- Parts
- Spares (loaners)
- Delivery
- Installation and pilot testing

- Predelivery inspection
- Problem diagnosis and repair
- User group

Quality

In this category, the quality and dependability of the various CAD/CAM systems is examined and compared. Vendors should be asked to present proof of the following:

- Corporate stability
- Satisfied customers
- Continued system improvement (research and development)
- Response to downtime problems
- Amount of downtime expected
- Quality of products produced on the system
- User group support
- Hardware reliability
- Software reliability
- Delivery and installation assistance

Delivery and Installation

In this category, the delivery and installation services provided are examined and compared. This is a critical area that must be closely coordinated between buyer and vendor. Criteria to look for include the following:

- Amount of time between purchase and delivery
- Predelivery inspection
- On-site setup
- On-site debugging and pilot testing
- Delivery of all supplies and interfacing mechanisms with system
- Postinstallation on training provided

GROUP TECHNOLOGY[1]

Group technology is a manufacturing concept in which similar parts are grouped together in parts groups or families. Parts may be alike in two ways:

---

[1]Adapted from David L. Goetsch, *Fundamentals of CIM Technology*, Chapter 6. Copyright 1988 by Delmar Publishers Inc.

1. design characteristics
2. manufacturing processes required to produce them

By grouping similar parts into families, manufacturing personnel can improve their efficiency. Such improvements are the result of advantages gained in such areas as setup time, standardization of processes, and scheduling.

Group technology can also improve the productivity of design personnel by decreasing the amount of work and time involved in designing new parts. A new part will likely be similar to an existing part in a given family. When this is the case, the new part can be developed by modifying the design of the existing part. Design modifications usually require less time and work than new designs. This is especially true in the age of CAD/CAM. Group technology is an important CAD/CAM concept in that it is a bridge builder between the design and manufacturing components.

Historical Background of Group Technology

Ever since the industrial revolution, manufacturing and engineering personnel have been searching for ways to optimize manufacturing processes. There have been numerous developments over the years since the industrial revolution mechanized production. Mass production and interchangeability of parts in the 1800s were major steps forward in optimizing manufacturing.

However, even with mass production and assembly lines, most manufacturing is done in small batches ranging from one workpiece to two or three thousand. Even today, over 70% of manufacturing involves batches of less than three thousand workpieces. Historically, less has been done to optimize small-batch production than has been done for assembly line work.

There have been attempts to standardize a variety of design tasks and some work in queueing and sequencing in manufacturing, but until recently, design and manufacturing in small-batch settings has been somewhat random.

The underlying problem that has historically prevented significant improvements to small-batch manufacturing is that any solution must apply broadly to general production processes and principles rather than to a specific product. This is a difficult problem because the various workpieces in a small batch can be so random and different.

When manufacturing entered the age of automation and computerization, such developments as scheduling software, sequencing software, and material requirements planning (MRP) systems became available to improve production of both small and large batches. Even these developments have not optimized the production of small-batch manufacturing lots. The problem has become even more critical because, since the end of World War II, the trend has been toward more small-batch and less large-batch production.

In recent years, the problems of small-batch production have finally begun to receive the attention necessary to bring about improvements. A major step is the ongoing development of group technology.

Part Families

It has already been stated that parts may be similar in design (size and shape) and/or in the manufacturing processes used to produce them. A group of such parts is called a part family. It is possible for parts in the same family to be very similar in design yet radically different in the area of production requirements. The opposite may also be true.

Figure 3–37 contains examples of two parts from the same family. These parts were placed in the same family based on design characteristics. They have exactly the same shape and size. However, they differ in the area of production processes. Part 1, after it is drilled, will go

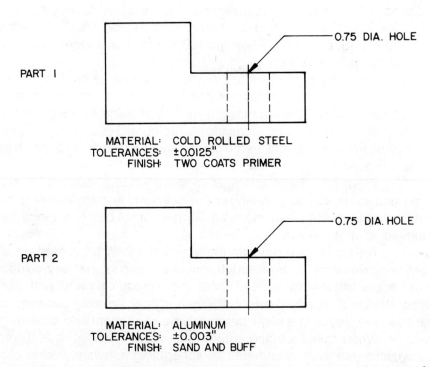

PART I

0.75 DIA. HOLE

MATERIAL: COLD ROLLED STEEL
TOLERANCES: ±0.0125"
FINISH: TWO COATS PRIMER

PART 2

0.75 DIA. HOLE

MATERIAL: ALUMINUM
TOLERANCES: ±0.003"
FINISH: SAND AND BUFF

FIGURE 3–37 Two parts from the same design part family. Reprinted from *Fundamentals of CIM Technology* by David Goetsch, copyright 1988 by Delmar Publishers Inc.

to a painting station for two coats of primer. Its dimensions must be held to a tolerance of ±0.125 in. Part 2, after it is drilled, will go to a finishing station for sanding and buffing. Its dimensions require more restrictive tolerances of ±0.003 in. The parts differ in material: Part 1 is cold-rolled steel and part 2 is aluminum.

Figure 3–38 contains examples of two parts from another family. Although the design characteristics of these two parts are drastically different (i.e., different sizes and shapes), a close examination reveals that they are similar in the area of production processes. Part 1 is made of stainless steel. Its dimensions must be held to a tolerance of ±0.002 in., and three holes must be drilled through it. Part 2, in spite of the differences in its size and shape, has exactly the same manufacturing characteristics.

PART I

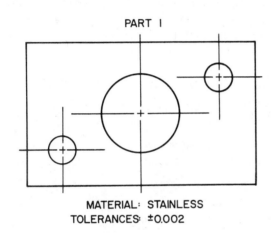

MATERIAL: STAINLESS
TOLERANCES: ±0.002

PART 2

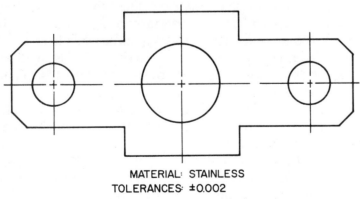

MATERIAL: STAINLESS
TOLERANCES: ±0.002

FIGURE 3–38 **Two parts from the same manufacturing part family.** Reprinted from *Fundamentals of CIM Technology* by David Goetsch, copyright 1988 by Delmar Publishers Inc.

The parts shown in Figure 3–37, because of their similar design characteristics, are grouped in the same family. Such a family is referred to as a **design part family**. Those in Figure 3–38 are grouped together because of similar manufacturing characteristics. Such a family is referred to as a **manufacturing part family**. The characteristics used in classifying parts are referred to as **attributes**.

By grouping parts into families, manufacturing personnel can cut down significantly on the amount of materials handled and movement wasted in producing them. This is because manufacturing machines can be correspondingly grouped into specialized work cells instead of the traditional arrangement of machines according to function (i.e., mills together, lathes together, drills together, etc.). This becomes very important when configuring flexible manufacturing cells.

Each work cell can be specially configured to produce a given family of parts. When this is done, the number of setups, the amount of material handling, the length of lead time, and the amount of in-process inventory are all reduced.

Grouping Parts into Families

Part grouping is not a simple process. You already know the criteria used: design similarities and manufacturing similarities. But how does the actual grouping take place?

Three methods can be used to group parts into families:

1. sight inspection
2. route sheet inspection
3. parts classification and coding

All three methods require the expertise of experienced manufacturing personnel.

Sight inspection is the simplest, least sophisticated method. It involves looking at parts, photos of parts, or drawings of parts. Through such an examination, experienced personnel are able to identify similar characteristics and group the parts accordingly. This is the easiest approach, especially for grouping parts by design attributes, but it is also the least accurate of the three methods.

The second method involves route sheets that are used to route the parts through the various operations to be performed. This can be an effective way to group parts into manufacturing part families, provided the route sheets are correct. If they are, this method is more accurate than the sight inspection approach. This method is sometimes referred to as the **production flow analysis (PFA) method**.

The most widely used method for grouping parts is the third method: **parts classification and coding**. This is also the most sophisticated, most difficult, and most time-consuming method. Parts classification and coding is complex enough to require a more in-depth treatment than the other two methods.

Parts Classification and Coding

Parts classification and coding is a method in which the various design and/or manufacturing characteristics of a part are identified, listed, and assigned a code number. Recall that these characteristics are referred to as attributes. This is a general approach used in classifying and coding parts. Many different systems have been developed to actually carry out the process, none of which has emerged as the standard.

The many different classification and coding systems that have been developed all fall into one of three groups:

1. design attribute group
2. manufacturing attribute group
3. combined attribute group

Design Attribute Group

Classification and coding systems that use design attributes as the qualifying criteria fall into this group. Commonly used design attributes include the following:

- dimensions
- tolerances
- shape
- finish
- material

Manufacturing Attribute Group

Classification and coding systems that use manufacturing attributes as the qualifying criteria fall into this group. Commonly used manufacturing attributes include the following:

- production processes
- operational sequence
- production time
- tools required
- fixtures required
- batch size

Combined Attribute Group

There are advantages in using design attributes and advantages in using manufacturing attributes. Systems that fall into the design attribute group are particularly advantageous if the goal is design retrieval. Those in the manufacturing group are better if the goal is a production-related function. However, there is a need for systems that combine the best characteristics of both. Such systems use both design and manufacturing attributes.

Sample Parts Classification and Coding System

Some companies develop their own parts classification and coding system; however, this can be an expensive and time-consuming approach. The more widely used approach is to purchase a commercially prepared system. There are several such systems available. However, the most widely used of these is the Opitz system. It is a good example of a parts classification and coding system and how one works.

Opitz System

Classification and coding systems use alphanumeric symbols to represent the various attributes of a part. One of the many classification and coding systems is the **Opitz system**. This system uses characters in 13 places to code the attributes of parts, and hence, to classify them. These digit places are represented as follows:

12345 6789   ABCD

The first five digits (12345) code the major design attributes of a part. The next four digits (6789) are for coding manufacturing-related attributes and are called the supplementary code. The letters (ABCD) code the production operation and sequence.

The alphanumeric characters shown above represent places. For example, the actual numeral used in each place can be 0 to 9. The numeral used in the "1" place indicates the length-to-diameter (L/D) ratio of the part. The numeral used in the "2" place indicates the external shape of the part. The numeral used in the "3" place indicates the internal shape of the part. The numeral used in the "4" place indicates the type of surface machining. The numeral used in the "5" place indicates the gear teeth and auxiliary holes. With such a system, a part might be coded as follows:

20801

The "2" means that the part has a certain L/D ratio. The first "0" means the part has no outstanding external shape elements. The "8" means the part has an internal thread. The second "0" means no surface machining is required. The "1" means the part is axial, not on pitch.

Figure 3–39 shows the basic overall structure of the Opitz system for parts classification and coding. Figure 3–40 is a chart used for assigning an Opitz code to rotational parts. Figure 3–41 is a drawing of an actual part that has been assigned an Opitz code of 15400. This code was arrived at as follows:

Step 1: The total length of the part is divided by the overall diameter:

$$\frac{1.75\ (L)}{1.25\ (D)} = 1.40$$

Since 1.40 is greater than 0.5 but less than 3, the first digit in the code is "1."

Step 2: An overall description of the external shape of the part would read ". . . . a rotational part that is stepped on both ends with one stepped end threaded." Consequently, the most appropriate second digit in the code is "5."

Step 3: A description of the internal shape of the part would read ". . . . a through hole." Consequently, the third digit in the code is "4."

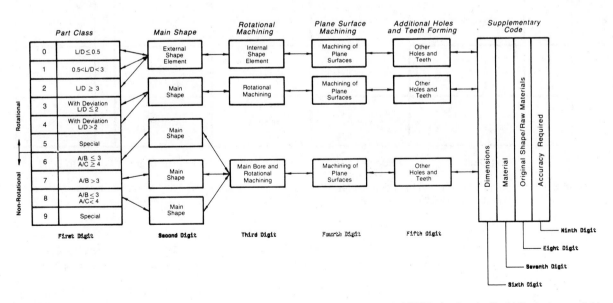

FIGURE 3–39 Basic structure of the Opitz system. Reprinted from *Fundamentals of CIM Technology* by David Goetsch, copyright 1988 by Delmar Publishers Inc.

PART CLASS		MAIN SHAPE EXTERNAL			ROTATIONAL MACHINING INTERNAL			PLANE SURFACE MACHINING	ADDITIONAL HOLES AND TEETH	
	L/D ≤ 0.5	Smooth, no shape elements			No hole, no breakthrough			No surface machining	No auxiliary hole	
1	0.5 < L/D < 3	Stepped to one end or smooth	No shape elements		Smooth or stepped to one end	No shape elements		Surface plane and/or curved in one direction, external	Axial, not on pitch circle diameter	No gear teeth
2	L/D ≥ 3		Thread			Thread		External plane surface related by graduation around a circle	Axial on pitch circle diameter	
3	—		Functional groove			Functional groove		External groove and/or slot	Radial, not on pitch circle diameter	
4	—	Stepped to both ends	No shape elements		Stepped to both ends	No shape elements		External spline (polygon)	Axial and/or radial and/or other direction	
5	—		Thread			Thread		External plane surface and/or slot, external spline	Axial and/or radial on PCD and/or other directions	
6	—		Functional groove			Functional groove		Internal plane surface and/or slot	Spur gear teeth	With gear teeth
7	—	Functional cone			Functional cone			Internal spline (ploygon)	Bevel gear teeth	
8	—	Operating thread			Operating thread			Internal and external polygon, groove and/or slot	Other gear teeth	
9	—	All others			All others			All others	All others	

First Digit   **Second Digit**   **Third Digit**   **Fourth Digit**   **Fifth Digit**

Rotational / Non-Rotational

FIGURE 3–40 Assigning a code using the Opitz system. Reprinted from *Fundamentals of CIM Technology* by David Goetsch, copyright 1988 by Delmar Publishers Inc.

Step 4: By examining the part, it can be seen that no surface machining is required. Therefore, the fourth digit in the code is "0."

Step 5: By examining the part, it can be seen that no auxiliary holes or gear teeth are required. Therefore, the fifth digit in the code is also "0."

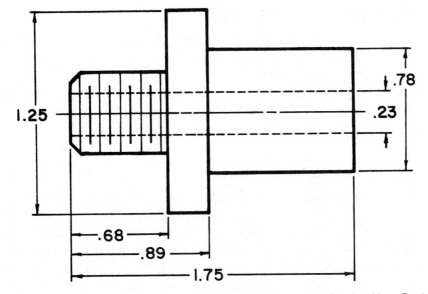

FIGURE 3-41 Sample part that can be coded using the Opitz system. Reprinted from *Fundamentals of CIM Technology* by David Goetsch, copyright 1988 by Delmar Publishers Inc.

ADVANTAGES AND DISADVANTAGES OF GROUP TECHNOLOGY

Several advantages can be realized through the application of group technology:

1. improved design
2. enhanced standardization
3. reduced materials handling
4. simplified production scheduling
5. improved quality control

Improved Design

Group technology allows designers to use their time more efficiently and productively by decreasing the amount of new design work required each time a part is to be designed. When a new part is needed, its various attributes can be listed. Then, an existing part with as many of these attributes as possible can be identified and retrieved. The only new design required is one that relates to attributes of the new part not contained in the existing part. Because this characteristic of group technology promotes design standardization, additional design benefits accrue.

Enhanced Standardization

Parts are classified into groups according to their similarities. The more similarities, the better. Consequently, enhanced standardization is promoted by group technology. As you have already seen, design factors account for part of the enhanced standardization. Setups and tooling also become more standardized since similar parts require similar setups and similar tooling.

Reduced Materials Handling

One of the major elements of group technology is the arrangement of machines into specialized work cells, each cell producing a given family of parts. Machines are traditionally arranged by function (i.e., lathes together in one group, mills together in another, etc.). A functional arrangement causes excessive movement of parts from machine to machine. Specialized work cells reduce such movement to a minimum and, in turn, reduce materials handling.

Simplified Production Scheduling

With machines grouped into specialized work cells and arrangement of parts into families, there is less work center scheduling required. There is also less scheduling within work cells. In addition to requiring less scheduling overall, group technology involves less sophisticated, less difficult scheduling.

Improved Quality Control

Quality control is improved through group technology because each work cell is responsible for a specific family of parts. This does two things, both of which improve the quality of parts: (1) it promotes pride in the work among employees assigned to each work cell, and (2) it makes production errors and problems easier to trace to a specific source. In traditional manufacturing shops, each work center performs certain tasks, but no one center is responsible for an entire part. Such an approach does not promote quality control.

In spite of these advantages, there are several disadvantages associated with group technology. They have to do with the transition from a traditional manufacturing approach to group technology. Rearranging the machines in a traditional shop into specialized work cells can cause a major disruption, which translates quickly into monetary losses.

As a consequence, some managers are reluctant to pursue group technology.

Another problem with group technology is that parts classification coding is a time-consuming, difficult, and expensive process, especially in the initial stages, and some managers balk at the cost.

KEY TERMS

CAD/CAM
Database
Interactive computer graphics
    (ICG)
Hardware
Software
Manual design
Response time
Automatic programmed tools
    (APT)
Sketchpad project
Turnkey system
Design modeling
Geometric model
Cathode ray tube (CRT)
Graphic image
Mathematical model
Orthographic representation
Oblique representation
Design analysis
Finite-element analysis
Design review
Semiautomatic dimensioning
Layering
Interference checking
Zoom capability
Kinematics
Design documentation
Transformations
Nonrepetition
Central processing unit (CPU)
Secondary storage

Workstation
Directed-beam refresh terminal
Refreshing the image
Direct-view storage tube
    (DVST) terminal
Raster scan terminal
Pixels
Resolution
Keyboard
Cursor control devices
Light pen
Electronic tablet
Digitizer
Windowing
Hidden line removal
Segments
User friendly
Wireframe models
Axonometric representation
Solid model
Constructive solid geometry
    (CGS)
Boundary representation (B-rep)
Animation
Applications checklist
Design part family
Manufacturing part family
Attributes
Production flow analysis
    (PFA) method
Parts classification and coding
Opitz system

Chapter Three REVIEW

1. Define CAD/CAM.
2. Explain the rationale for CAD/CAM.
3. What is the APT language, where did it come from, and why was it developed?
4. Why was the Sketchpad Project significant?
5. Why did early microcomputer-based CAD/CAM systems fail?
6. Explain the following:
   Design modeling
   Design analysis
   Design review
   Design documentation
7. What is finite-element analysis?
8. Explain the "threw the plans over the wall" problem.
9. What is the CAD/CAM database and what does it contain?
10. What are the components of a typical CAD/CAM hardware configuration?
11. Explain the three main types of CAD/CAM terminals. How does each create a graphic image?
12. What is a digitizer?
13. What are the most widely used cursor control devices?
14. Compare the relative merits of 2-D and 3-D software.
15. Compare the relative merits of 3-D wireframe models and solid models.

These case studies were provided by the Society of Manufacturing Engineers (SME). They are excerpted from the SME's *Manufacturing Insights*® series of videotapes. These case studies give students actual examples of the way the advanced manufacturing technologies covered in this chapter are being applied in "real-world" manufacturing settings.

As you read each case study, relate the examples cited to the material presented in the text of the chapter. This combination of text-book information and real-world examples will be particularly valuable to your understanding of advanced manufacturing technologies.

K2 SKI COMPANY

The K2 Ski Company of Vashon Island, Washington, makers of some of the most popular and successful skis in the industry, uses CAD/CAM to achieve dramatic savings. At K2, the use of CAD/CAM begins in the design department, where the CAD terminals are used to design product configurations, forms, fixtures, and hard dies that are used in the production process.

In the case of dies and fixtures, prints are made and sent to the tool room, where numerical control (NC) machines are used to cut the fixtures and dies, which are then finished and supplied to the production floor. The hard tooling and dies used throughout the assembly process are produced here.

The assembly of a ski begins with the cutting and forming of the steel edge material, which is then cemented to the pitex form and is ready for molding. Two cores are used at K2: special laminated wood and variable-density durafoam that is heavier at the tip and back for greater break strength. K2 also weaves their own fiberglass to control quality and specifications to meet their rigid requirements.

All of these elements come together in the mold room, the most critical step in the process, where the fiberglass layers are wetted down with epoxy resin and wrapped in layers around the core. This layup is then put into a mold, the base assembly put on top, and the mold closed to press the assembly at 400°F until the resin cures. The skis are sanded to remove any excess material, then they go through a series of wet sanders for the top, bottom, and sides. Finally, the bases receive a special stone grinding and polishing. The silkscreened tops are then glued on to finish the assembly process.

(continued)

The edges of the skis are then beveled and matched into pairs that have the same camber and flex. Serial numbers are added to the pairs, the skis receive a final cleaning and waxing, the steel edges are lightly oiled, and they are shrink-wrapped for shipping.

The K2 Ski Company's use of computer-aided design has helped them build an extensive database that is available for reference and review at all times by their designers and consulting experts. Their computer-aided manufacturing program helps them produce top-quality tooling, dies and fixtures on site, providing the precision quality they require, on the assured delivery schedule they need. Throughout the assembly process, bar code readers provide computer input on production flow and material requirements to their materials requirements planning system.

This combination of a CAD/CAM and a MRP system helps K2 shorten the development and production time of new products. It also allows them to take advantage of the knowledge gained in previous designs and programs to constantly improve their products and maintain their competitive position in the rapidly growing leisure products and sports equipment field.

FLOW SYSTEMS, INC.

Flow Systems, Inc., of Kent, Washington, makes excellent use of an expanding design database from their CAD system. Flow Systems manufactures a wide variety of waterjet cutting systems for both nonmetallic and metallic applications. Because of the variety of requirements, Flow Systems produces a line of standard, "off-the-shelf" cutting machines and designs and produces a large number of custom systems for specific applications.

The development of new standard products and custom design systems begins at a CAD terminal, using the existing database as a starting point. The special features to be incorporated are then added and a new design created.

From the CAD department, the design engineer electronically sends the cutter path program for the parts directly to the computer numerical control (CNC) machines on the shop floor. There the machine operator reads the program off the screen, loads the designated tools,

Image selected from SME's CAD/CAM video from the Manufacturing Insights® Video Series.

clamps the stock in place and starts the machine cutting the part for assembly into the new system.

In addition to prototype parts, standard units and quantity production of special order units is also handled by numerical control (NC) and CNC machines in the production area, which are controlled by the CAM system to provide precise manufacturing control, production, and inventory data.

Flow Systems' established database helps them to reduce design time for new and special products and reduces the time to production of new designs. This provides cost savings in development and production and allows them to provide improved customer service and delivery. It is also a valuable sales tool in providing customer inquiry response.

Dr. Gordon Kirkpatrick of Flow Systems, Inc. states:

One of the things we commonly do is a lot of custom cutting of samples. If a customer comes in and wants a sample part cut,

(continued)

he frequently gives us a sketch. We go to a CAD terminal in engineering and we draw his part in front of him, full size in three dimensions. At the same time, the manufacturing engineer, sees the part appear on the screen. He has NC capability, so as fast as it appears, he generates an NC tape, which we feed to a robot in the demonstration room. By the time the customer gets downstairs, he can see the robot cutting out the part. We use that not just as a manufacturing gimmick, but as a prototype for what we expect to happen out here in the future.

In addition to instantaneous prototyping, Flow Systems has found many uses for the common database they have established for their operations. Dr. Kirkpatrick states:

We create one master database, which consists of the product assembled in three dimensions. We use this for flat pattern generation; for tooling design—we design the tools around the master image; NC tape production; technical illustration manuals; demonstrations for the customer; not so much for pure show but to make sure we're building what the customer wants; and we expect to use it in the future by putting read-only terminals on the shop floor to allow people who actually run the machines to both summon up drawings they need and comment on designs we're making.

At Flow Systems, Inc., CAD/CAM is making a major contribution to the flexibility, profitability and growth of this progressive and innovative small company.

TORO OUTDOOR PRODUCTS

The Toro Outdoor Products Group plant in Minneapolis, Minnesota, has found CAD/CAM to be very practical and useful in the improvement of existing products and development of new ones. At Toro, the CAD department is the starting place for new product designs and engineering. A large database has been developed that contains extensive information, details, and drawings on current products. This information

Image selected from SME's CAD/CAM video from the Manufacturing Insights® Video Series.

is used as the basis for developing new products as well as design changes and modifications to improve existing products.

For example, CAD was used in developing a new chute design for the grass catcher attachment on a small lawn tractor. This design is based on the original part, since it has to fit the same mounting fittings and attachments, and it incorporates the new interior shape to facilitate a smoother flow of clippings.

When the design parameters are agreed on, the same database is used to generate cutter path instructions for the CNC machine that will cut the prototype. When the cutter path data are complete and have been checked for accuracy and fixture clearance, it is downloaded to the CNC machine, where the actual prototype part is machined out of a block of epoxy composite material. It can then be finished, painted, and tried on the tractor in actual tests to determine strength and performance.

(continued)

Dana Lonn, Chief Engineer at Toro Outdoor Products states:

We think there are a number of benefits, including reduced development time, improved quality, parts fit together a lot better, we're able to more thoroughly analyze parts, we're able to build much more complex parts much more accurately for prototype parts, and for production parts. Once we get a part that's acceptable, we can reproduce an exact copy of that part.

We think the maximum benefit is gained by doing multiple applications where you can reuse the same geometry over and over again. That's where the productivity is in computer-aided design and computer-aided manufacturing. You don't save any time in drawing the part the first time. It's when you use that same part over and over. You can use it for your technical publications, you can use it to build NC programs, you can use it to do another design, you can use it for assembly drawings.

We know a product we produced last winter called a CCR 2000 rotor. That specific rotor, the tooling and the checking fixtures and the prototype parts were built off CAM models and we know that we saved in excess of fifty thousand dollars in producing that part and the tooling.

CAD/CAM has been highly successful at Toro Outdoor Products, where they are continuing to expand their system to other parts of their operations. At the same time, they are adding new capabilities to their present installation to provide greater flexibility and capacity.

OSTER DIVISION OF SUNBEAM CORPORATION

The Oster division of Sunbeam Corporation, in Milwaukee, Wisconsin, maker of home convenience and comfort appliances, uses CAD in the design of a number of consumer products. At Oster, product design changes and improvements and new product designs, originate in the engineering department, at one of several CAD terminals. Designs are developed, analyzed, and simulated on computers, using design data accumulated in the master database, before committing them to prototype production.

After the operating parts of the design are complete, an envelope is designed around the mechanism to provide a pleasing and attractive, as well as functional, case. New product designs are reviewed with both marketing and manufacturing while still in the computer 3-D solid model stage to assure marketability and producibility.

Once a design has been agreed on, an NC cutter path program is generated and sent to the prototype shop, where an initial model of the case is cut on the machine from a block of wood. The cutter path program used to make this original model can then be used as the basis for the final program used to produce the production mold dies. When the model is cut and finished, it is painted to represent the final product, and then used for final marketing and production review before a working prototype is made.

At Oster, CAD/CAM is a key element in the company's continuing product development and improvement program, helping them to reduce product prototyping and development costs and reducing the time required to get new products to market, and helping Oster remain a leading producer of small appliances and convenience products for the home in an increasingly competitive market.

Major Topics Covered

- Overview of Numerical Control
- Historical Development of NC
- NC Machine Components
- NC Programming
- Classifications of NC Machines
- Pros and Cons in Justifying NC
- Pros of an NC Conversion
- Cons of an NC Conversion
- MDI Control
- Computer-Assisted NC Programming
- Benefits and Gains from NC
- Overview of Direct Numerical Control
- Historical Development of Direct Numerical Control
- Direct, Computer, and Distributed Numerical Control
- Data Transmission in DNC
- Advantages of DNC
- Best Applications of DNC
- Overview of CNC
- Elements of a CNC System
- Historical Development of CNC
- CNC Versus Direct Numerical Control
- Operator Interfaces in CNC
- *Case Study: Lockheed–Georgia*
- *Case Study: Badger Meter*

Chapter Four

Automated Assembly

Automated Guided Vehicles

CAD/CAM

Numerical Control ◄

Industrial Robots

Lasers in Manufacturing

Programmable Logic Controllers

Flexible Manufacturing

Computer Integrated Manufacturing

Other Related Technologies

OVERVIEW OF NUMERICAL CONTROL

One of the most fundamental concepts in the area of advanced manufacturing technologies is **numerical control (NC)**. Prior to the advent of NC, all machine tools were manually operated and controlled. Among the many limitations associated with manual control machine tools, perhaps none is more prominent than the limitation of operator skills. With manual control, the quality of the product are directly related to and limited to the skills of the operator. Numerical control represented the first major step away from human control of machine tools.

Numerical control means the control of machine tools and other manufacturing systems through the use of prerecorded, written symbolic instructions. Rather than operating a machine tool, an NC technician writes a program that issues operational instructions to the machine tool. For a machine tool to be numerically controlled, it must be interfaced with a device for accepting and decoding the programmed instructions, known as a **reader**.

Numerical control was developed to overcome the limitation of human operators, and it has done so. Numerical control machines are more accurate than manually operated machines, they can produce parts more uniformly, they are faster, and the long-run tooling costs are lower. The development of NC led to the development of several other innovations in manufacturing technology:

- Electric discharge machining
- Laser cutting
- Electron beam welding

Numerical control has also made machine tools more versatile than their manually operated predecessors. An NC machine tool can automatically produce a wide variety of parts, each involving an assortment of widely varied and complex machining processes. Numerical control has allowed manufacturers to undertake the production of products that would not have been feasible from an economic perspective using manually controlled machine tools and processes.

HISTORICAL DEVELOPMENT OF NC

Like so many advanced technologies, NC was born in the laboratories of the Massachusetts Institute of Technology. The concept of NC was developed in the early 1950s with funding provided by the U.S. Air Force. In its earliest stages, NC machines were able to make straight cuts efficiently and effectively.

However, curved paths were a problem because the machine tool had to be programmed to undertake a series of horizontal and vertical steps to produce a curve. The shorter the straight lines making up the steps, the smoother is the curve (Figure 4–1). Each line segment in the steps shown in the closeup in Figure 4–1 had to be calculated. This was a cumbersome approach that had to be overcome if NC was to develop further.

This problem led to the development in 1959 of the **Automatically Programmed Tools (APT) language**. This is a special programming language for NC that uses statements similar to English language to define the part geometry, describe the cutting tool configuration, and specify the necessary motions. The development of the APT language was a major step forward in the further development of NC technology. The original NC systems were vastly different from those used today. The machines had hardwired logic circuits. The instructional programs were written on punched paper, which was later to be replaced by magnetic

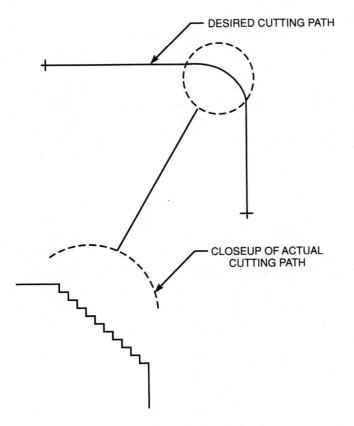

FIGURE 4–1 Machine tool cutting path.

plastic tape. A tape reader was used to interpret the instructions written on the tape for the machine. Together, all of this represented a giant step forward in the control of machine tools. However, there were a number of problems with NC at this point in its development.

A major problem was the fragility of the punched paper tape medium. It was common for the paper tape containing the programmed instructions to break or tear during a machining process. This problem was exacerbated by the fact that each successive time a part was produced on a machine tool, the paper tape carrying the programmed instructions had to be rerun through the reader. If it was necessary to produce 100 copies of a given part, it was also necessary to run the paper tape through the reader 100 separate times. Fragile paper tapes simply could not withstand the rigors of a shop floor environment and this kind of repeated use.

This led to the development of a special magnetic plastic tape. Whereas the paper tape carried the programmed instructions as a series of holes punched in the tape, the plastic tape carried the instructions as a series of magnetic dots. The plastic tape was much stronger than the paper tape, which solved the problem of frequent tearing and breakage. However, it still left two other problems.

The most important of these was that it was difficult or impossible to change the instructions entered on the tape. To make even the most minor adjustments in a program of instructions, it was necessary to interrupt machining operations and make a new tape. It was also still necessary to run the tape through the reader as many times as there were parts to be produced. Fortunately, computer technology became a reality and soon solved the problems of NC associated with punched paper and plastic tape.

Advent of Direct Numerical Control

The development of a concept known as **direct numerical control (DNC)** solved the paper and plastic tape problems associated with numerical control by simply eliminating tape as the medium for carrying the programmed instructions. In direct numerical control, machine tools are tied, via a data transmission link, to a **host computer** (Figure 4–2). Programs for operating the machine tools are stored in the host computer and fed to the machine tool as needed via the data transmission linkage. Direct numerical control represented a major step forward over punched tape and plastic tape. However, it is subject to the same limitations as all technologies that depend on a host computer. When the host computer goes down, as happens with all host computers, the machine

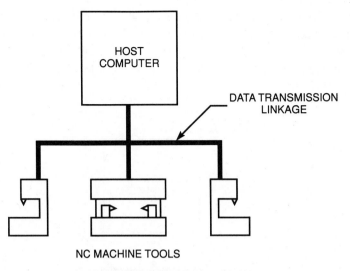

FIGURE 4–2 Direct numerical control.

tools also experience downtime. This problem led to the development of computer numerical control.

Advent of Computer Numerical Control

The development of the microprocessor allowed for the development of **programmable logic controllers (PLCs)** and microcomputers. These two technologies allowed for the development of **computer numerical control (CNC)**. With CNC, each machine tool has a PLC or a microcomputer that serves the same purpose. This allows programs to be input and stored at each individual machine tool. It also allows programs to be developed off-line and downloaded at the individual machine tool. CNC solved the problems associated with downtime of the host computer, but it introduced another problem known as **data management**. This is a problem all work settings dependent on microcomputers have. The same program might be loaded on ten different microcomputers with no communication among them. This problem is in the process of being solved by local area networks that connect microcomputers for better data management. The problem of data management led to the development of distributed numerical control.

Advent of Distributed Numerical Control

Distributed numerical control (also called **DNC**) takes advantage of the best aspects of direct numerical control and computer numerical control. With distributed numerical control there are both host computers and local computers at the individual machine tools (Figure 4–3). This allows the programs to be stored in the host computer and, thereby, better managed. However, it also allows them to be downloaded to local microcomputers or PLCs. It also allows for local input and interaction through microcomputers or PLCs at the machine level.

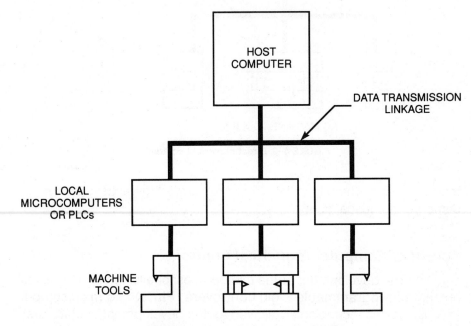

FIGURE 4–3 Distributed numerical control.

NC MACHINE COMPONENTS

There are four components in any NC machine (Figure 4–4):

1. the actual NC tool
2. the **machine control unit (MCU)**
3. the **communication interface** between the NC machine and the MCU
4. a variety of accessories for performing specific jobs on the NC machine (Figure 4–4)

The actual NC machine may be a milling machine, lathe, drill, or any other type of machine tool. The MCU is the control unit that holds the programs that instruct the NC machine. The MCU also has various devices available for operator input. Information contained in the MCU is carried to the

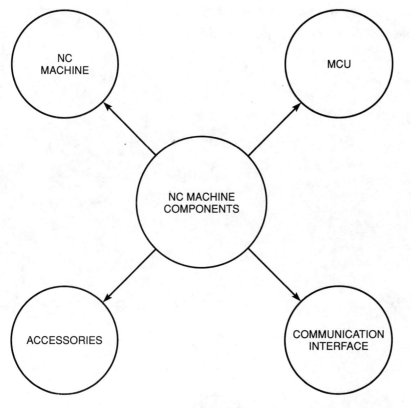

FIGURE 4–4 NC machine components.

activators on the NC machine through the communication interface. These activators receive the electronic signals from the MCU and cause the mechanical apparatus of the NC machine to operate.

Less sophisticated NC machines have **open-loop activators**. An open-loop activator can receive a signal and carry out the instructions contained in that signal, but cannot feed back to the MCU to show that the instructions carried in the signal have been properly completed. More sophisticated NC machines use **closed-loop activators**. A closed-loop activator can receive and carry out a signal and feed data back to the MCU showing that the signal has been carried out and to what extent. The more sophisticated closed-loop systems are being used more and more because they allow for closer monitoring and immediate corrective action when problems with executing a program arise. The accessories are special tools required to carry out a specific job. Figures 4–5 and 4–6 show NC machines and typical components.

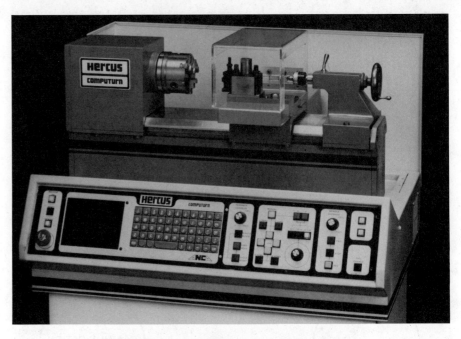

FIGURE 4–5 CNC lathe. *Courtesy of Computurn.*

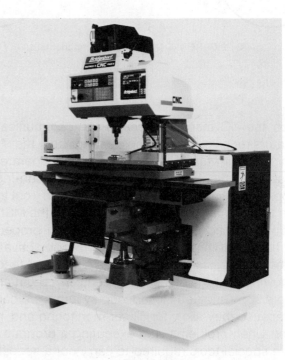

FIGURE 4–6 CNC milling machine. *Courtesy of Bridgeport-Textron.*

NC PROGRAMMING

There are four ways to program an NC machine: **manual programming, digitizing, written programs,** and **graphic programs** (Figure 4–7). Manual programming is the most cumbersome of the four. It involves calculating numerical values that identify tool location and specify tool directions. Once these values have been calculated, they are recorded and fed into the MCU. Digitizing is a process frequently used in computer-aided design and drafting, whereby a drawing of a part is traced electronically. As it is traced, the various points on the drawing are converted into X-Y coordinates and stored in the computer. Once the drawing has been completely traced, the stored X-Y coordinates define the part and can be fed to an NC machine to provide instructions on tool positioning and movement.

Written programs are similar to those developed for use with any computer. With such programs, English language-type statements are written to describe tool positions and movement, as well as speed and feed rates. Such programs are fed into the MCU, where they are

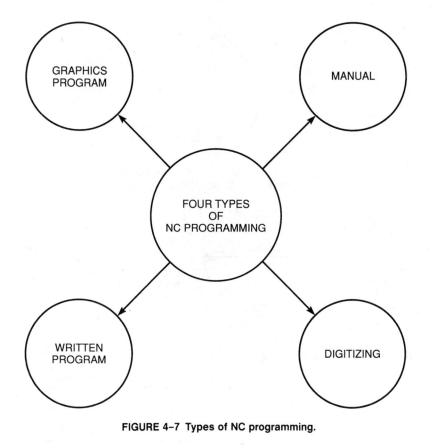

FIGURE 4–7 Types of NC programming.

translated into machine language and forwarded to the NC machine's activators.

The most modern, sophisticated method of programming an NC machine is by using a three-dimensional model of the part to provide the data that guide the NC machine in producing the part. As NC technology continues to develop, this programming method will eventually be used more than any other.

CLASSIFICATIONS OF NC MACHINES

Numerical control machines are classified in different ways. An early method was to categorize them as being either **point-to-point** or **continuous-path** machines. Point-to-point machines, as the name implies, move in a series of steps from one point to the next (Figure 4–8).

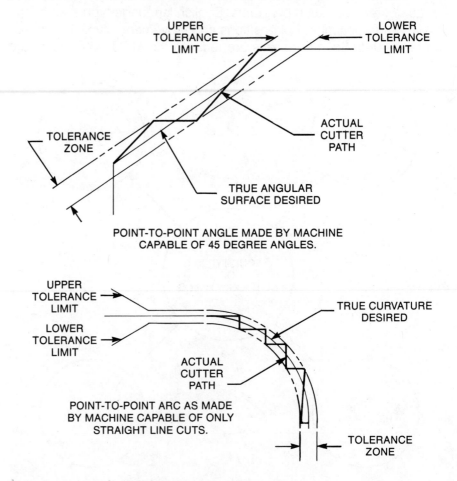

FIGURE 4–8 **Point-to-point cutter path.** Reprinted from Seames/*Computer Numerical Control: Concepts and Programming*, copyright 1990 by Delmar Publishers Inc.

Point-to-point machines are less sophisticated and less precise than continuous-path machines. Continuous-path machines move uniformly and evenly along the cutting path rather than through a series of horizontal and vertical steps. Such machines are more sophisticated and require more memory in the MCU than point-to-point machines. Figure 4–9 illustrates the types of cutting paths performed by continuous-path machines.

Another way to classify NC machines is as either **positioning** or **contouring** machines. Point-to-point machines are considered positioning machines. Continuous-path machines are considered contouring machines. Positioning machines have as few as two axes: the X axis, and the Y axis. Contouring machines must have at least three axes: the X, Y, and Z axes. Figure 4–10 illustrates the movements governed by the X, Y, and Z axes. Note that X represents the **longitudinal** axes, Y the **transverse axis**, and Z the up-and-down or **vertical axis**. Figure 4–11 is a simple line diagram of a typical three-axis machine tool showing how movement is accomplished. On some machines, movement is accomplished

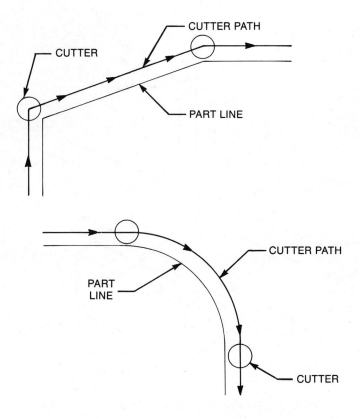

FIGURE 4–9 Continuous path cutter path. Reprinted from Seames/*Computer Numerical Control: Concepts and Programming*; copyright 1990 by Delmar Publishers Inc.

KALAMAZOO VALLEY COMMUNITY COLLEGE LIBRARY

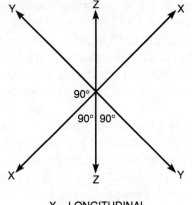

X = LONGITUDINAL
Y = TRANSVERSE
Z = VERTICAL

FIGURE 4–10 *X-Y-Z* axes.

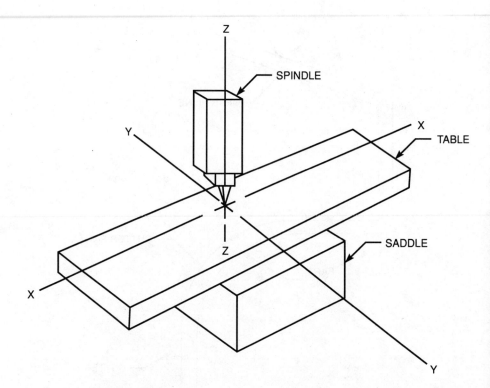

FIGURE 4–11 How the *X-Y-Z* axes relate to a CNC machine tool.

by positioning the spindle, and thus the tool, longitudinally along the X axis, transversely along the Y axis, and vertically along the Z axis. The workpiece is affixed to the table. With other machines, both the spindle and the table (thus the workpiece) can be moved.

Positioning machines work well for drilling applications. Milling operations are more likely to be contouring machines to allow for three-dimensional control.

Some of the more sophisticated positioning machines are able to accomplish angular cuts known as **slopes**. These are cuts that move across the quadrants formed by the intersection of the X and Y axes at angles other than 90 deg to either the X or Y axis (Figure 4–12). Slopes are generally imprecise and inaccurate. However, there are instances in which the ability to make angular cutting paths is important. In these cases, slopes can be an important feature, particularly where the cut surfaces do not have to mate with another surface. When precise, accurate angular cutting paths must be made, a contouring machine is needed.

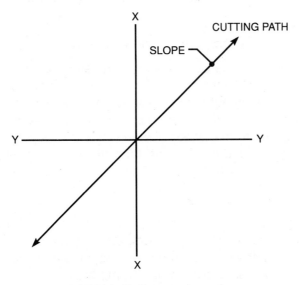

FIGURE 4–12 Slope cutting path.

PROS AND CONS IN JUSTIFYING NC

Justifying a conversion from manual to NC machines and processes can be difficult. From a strictly business perspective, such a conversion entails a great deal of risk, primarily centering around a perceived large capital outlay expenditure to purchase the necessary equipment or to retrofit existing equipment. There is also the human element to be considered: How will machining personnel react to the conversion? Will they be able to successfully make the conversion and operate NC machines?

Numerical control will ultimately succeed or fail depending on how the conversion is handled. The critical question that must be answered is, "Should there even be a conversion in the first place?" A five-step process based on the following questions can help manufacturing personnel find the correct answer before attempting a conversion from manual machines and processes to NC:

1. Why is a conversion being considered?
2. What are the specific objectives of the conversions?
3. What are all of the various different ways these objectives can be met?
4. What are the projected costs versus gains for each alternative?
5. How does the human factor affect each alternative?

Why Is a Conversion Being Considered?

Whenever a new technology begins to appear in the literature and sales representatives start calling to sell their latest systems, it is only natural for manufacturing personnel to begin wondering if they are falling behind the times and if, perhaps, it is time to make a change. However, a conversion from manual machines and processes to NC involves too much expense and risk to undertake without first completing a thorough examination of the situation. This examination should include the answer to the question, "Why is a conversion being considered?"

What does the company expect to gain from a conversion? Does the company wish to expand into a new product market? Is the company trying to become more competitive in the marketplace? Is the company simply trying to hold its own against increased competition from other companies who have already made a conversion? Regardless of the reason for considering a conversion, it is important to begin with a clear understanding of that reason.

What Are the Objectives of the Conversion?

Once the reason for considering a conversion has been defined, the next step is to frame that reason with specific objectives. These objectives should be stated in measurable terms so that there is a definite answer as to whether the conversion or any other alternative will meet the objectives. Is the company trying to improve overall productivity? Is the company trying to cut down on the amount of waste? Is the company trying to decrease its direct labor costs? There are numerous specific objectives a company might wish to accomplish through a conversion to NC. Whatever these objectives are, they should be identified and stated.

What Are the Alternatives?

Once a list of specific objectives has been drafted, manufacturing personnel can explore a variety of options for accomplishing them. Are there ways to accomplish the objectives other than converting to NC? One alternative to consider could be continuing in the existing format. Another alternative could be a complete conversion to NC. Other alternatives might involve a partial conversion to NC for certain products and processes and a continuation of the existing format for others. Another alternative might be to convert to NC while purchasing all new NC machines and equipment. Yet another alternative could be to retrofit existing machines and equipment so that they could be numerically controlled. Every possible alternative for accomplishing the stated objective should be listed.

What Are the Projected Costs Versus Gains?

This is a critical question. For every possible alternative, the projected gains, the projected costs, and the projected risks must be identified. What are the gains in terms of direct labor costs? What are the gains in terms of tool and fixture costs? What are the gains in terms of inspection costs? What are the gains in terms of maintenance costs? What are the costs in terms of initial capital outlay expenditures? What are the indirect costs? What are the risks associated with each option? These types of questions must be asked and answered for every possible alternative. Once the gains, costs, and potential risks for each option have been recorded, judgments can be made as to what option represents the best alternative for the specific company. However, before making those judgments, there is a final step to complete.

What About the Human Factor?

In addition to the costs, gains, and potential risks of each identified option the human factor must be considered. Will current employees be able to make whatever transition is associated with each alternative? Is training available where it will be needed? If a union is present, what is its attitude toward the various alternatives?

Once the costs, gains, risks, and human factors for each specific alternative have been considered, manufacturing personnel have the information they need to recommend the best alternative. If the recommendation is to proceed with a conversion to NC, whether that means the retrofitting option or the new machines option, there will be a list of predictable pros and cons associated with the conversion.

PROS OF AN NC CONVERSION

Enough companies large and small have undertaken the conversion from manual machines and processes to NC that an experience base has been created, which shows that certain types of benefits usually result from a conversion to NC. These benefits are generally found in the following areas:

- Direct labor savings
- Tool and fixture cost savings
- Consumable tool cost savings
- Inventory cost savings
- Tool setting cost savings
- Inspection cost savings
- Maintenance cost savings
- More consistent part quality
- Permanent memory of how a part was made

CONS OF AN NC CONVERSION

In spite of the potential benefits of an NC conversion, there are negative aspects as well. The most prominent are the following:

- Large initial expenditures on equipment or retrofitting
- Programming costs
- Training or retraining costs for the existing work force
- Potential rejection of the conversion by experienced personnel
- Potential union problems
- Higher consumable tool costs per hour of operation

The last item, higher consumable tool costs, is an often-stated disadvantage of NC. Those who do not support NC as a concept often use the higher consumable tool cost figures as the basis for their argument. However, to simply say that the consumable tool costs for NC are higher per hour of operation is to tell only half of the story. The tool costs per hour of operation are higher with NC, but this is because more metal is actually removed by the tools during the same period of time. Hence, the tools are doing more work per hour of operation. In other words, the manually operated machines use fewer consumable tools per hour of operation, but they also require more hours to produce the same amount of work. Therefore, in the long run, the overall tool consumption costs for NC are actually the same as for manual machines for a given run length.

MDI CONTROL

One of the limitations of traditional NC is an inability to program a machine on the shop floor. This problem is solved to a great extent by **manual data input (MDI) control**. With MDI control, an operator can enter data via a computer at the machine on the shop floor. This allows the operator to do a limited amount of programming at the machining site. The hardware for MDI control is small and not obtrusive. Manual data input is easy to use and does not require the skills of a professional part programmer. The operator enters the programming data via keyboard and monitors his entries by watching a CRT terminal. Manual data input control is so simple that it amounts to little more than an operator viewing a terminal display and filling in the blanks as necessary.

Manual data input control can be used with a wide range of machines from-low level machines to highly sophisticated machining centers. However, the most common uses are the less sophisticated three-axis milling machines and two-axis lathes.

COMPUTER-ASSISTED NC PROGRAMMING

The computer is becoming widely used in NC programming settings. Computers offer operators several advantages over manually preparing programmed instructions for NC machines. It takes less time to prepare a part program using a computer than to prepare the same program manually. Because the computer performs any necessary calculations rather than the calculations being performed manually, there are fewer errors in the final program. And, because less time is required in using a computer and fewer errors appear in the final program, the overall programming costs are usually lower with computer-assisted NC programming.

When using a computer to write an NC program, the operator describes the operations to be performed by the machine tool using English language-like commands. These commands are transformed into a language the NC machine can understand by a **post-processor**, a special computer program that converts general instructions to machine-specific instructions. The most widely used language for computer-assisted NC programming is the APT language.

The computer offers the NC part programmer the same advantages and benefits it offers other technical workers. By performing complicated mathematical calculations, it reduces the time involved in preparing a program and decreases the number of errors made in producing a program. When errors are made, they are easier to correct. It simplifies the input of programs to be stored, retrieved, and used continually without having to rerun a tape each time a machine operation is to be performed.

BENEFITS AND GAINS FROM NC

Numerical control has been in use for almost 30 years. Since its inception, it has been improved continually. Each improvement has added a new benefit or improved an existing benefit of NC as compared with traditional manual machine operation. During this period, a body of knowledge has developed of the actual benefits that can be derived from NC. There is a general consensus among manufacturing professionals that the principal benefits derived from NC are the following:

- Better planning
- Greater flexibility
- Easier scheduling
- Less setup, lead, and processing time
- Better machine utilization
- Lower overall tooling costs
- Greater uniformity in cutting
- Greater accuracy in cutting
- A higher degree of interchangeability of parts and tools
- More accurate cost estimates
- Permanent memory of how a part was made

These are the same types of benefits generally associated with any manufacturing process that has successfully moved from a manual format to a fully or partially automated format. The extent of the benefit depends on how successfully the transition has been carried out and how well developed the associated technologies have become. With the advent of DNC and CNC in numerical control, all of the benefits listed above have evolved into bona fide, documentable benefits.

OVERVIEW OF DIRECT NUMERICAL CONTROL

Direct numerical control is an advanced form of numerical control. Traditional NC involves imprinting programmed instructions on punched paper tape or magnetic plastic tape and feeding the tape through a reader that then interprets the instructions and transmits them to the NC machine. With direct numerical control, the tape media are eliminated. Instead, NC machines are connected to a host computer via a data transmission interface (Figure 4–13). Programs are stored in the host computer and downloaded to the individual NC machines via the data transmission interface as needed.

In addition to eliminating the cumbersome tape medium for carrying programmed instructions, direct numerical control allows NC machines at remote locations to be connected to the host computer. It also allows programmers to develop programs at any location and enter them into the host computer. This means that programmers are no longer physically tied to the NC machine and NC machines are not tied down with the traditional tape media.

HISTORICAL DEVELOPMENT OF DIRECT NUMERICAL CONTROL

Direct numerical control originated in the mid 1960s. Its original purpose was to reduce the amount of hardware required to provide NC. One host computer could serve as the controller instead of having a controller for each individual NC machine. The requirement of having one controller per NC machine was an expensive requirement in the mid-1960s because microelectronic technology had not yet developed to the point where it could be as inexpensive as it is today. As microelectronic technology and computer technology continued to evolve over the years, becoming more sophisticated but less expensive, the original rationale for direct numerical control has become less critical.

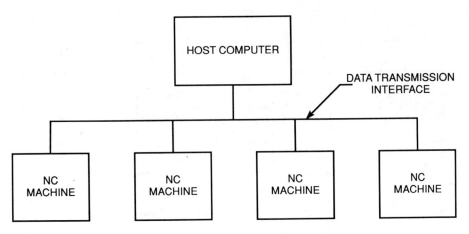

FIGURE 4–13 Direct numerical control.

Direct numerical control never worked completely as it was originally envisioned. Elimination of tape as the input medium and of controllers at each individual NC machine were never fully realized because of breakdowns in the data transmission interface and downtime of the host computer. When the host computer is down, it is necessary to have individual controllers at each NC machine and tape to input the instructions on a backup basis. It was only after the advent of computer numerical control (CNC) that the true value of direct numerical control began to be realized.

DIRECT COMPUTER AND DISTRIBUTED NUMERICAL CONTROL

Distributed numerical control is a concept that combines the best of direct and computer numerical control. The advent of CNC allowed the placement of a microcomputer controller at each individual NC machine. This computer could be used to develop programs, store programs, and input programs to the NC machines. This capability, coupled with the host computer capability of direct numerical control, paved the way for the distributed numerical control concept. In distributed numerical control, there is a host computer as in direct numerical control, as well as microcomputer controllers for individual NC machines (Figure 4–14).

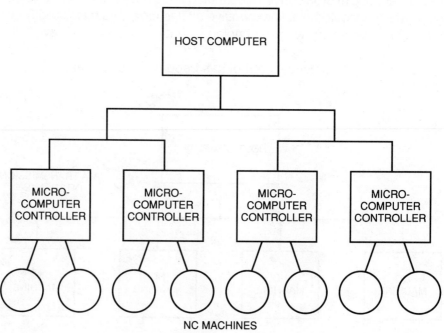

FIGURE 4–14 Distributed numerical control.

The host computer in such a system is still used as the main storage point for programs. These programs are downloaded from the host computer to the microcomputers, where they can be stored or transmitted to the NC machines. In this way, the need for the tape medium and reader is eliminated. The microcomputer controllers can also serve as backup memory when the host computer is down. Therefore, the NC machines do not have to be down when the host computer is down. This concept of distributed control, which combines direct and computer numerical control, is becoming so widely accepted that the acronym "DNC" is being used more and more to mean distributed numerical control instead of direct numerical control.

DATA TRANSMISSION IN DNC

Dependable data transmission is the key to successful DNC. A poor data transmission network can cause NC machines to be idle while waiting for instructions to be transmitted from the host computer. This can also result in operators waiting for NC machines to respond to the instructions from the host computer. Therefore, it is critical to have an effective data transmission network from the host computer to the intermediate microcomputer controllers and to the NC machines. One key to having an effective data transmission network is to ensure that the data transmission interface is compatible, not just with the host computer, but with each individual microcomputer controller and the NC machines.

This means the host computer must be able to feed the intermediate microcomputer controllers only as much of a program as they are able to hold in storage. The DNC system must also be able to download only those portions of a program that a given NC machine is capable of accepting and carrying out. The system should be able to accommodate revisions to programs in order to optimize the program of a given part without replacing the original program with the optimized program.

ADVANTAGES OF DNC

One key advantage of DNC, is the ability to produce and print reports that provide valuable information to system managers. This information can be studied and used to improve the performance of an overall DNC system. The types of reports that can be produced from a DNC system include the following:

- Production schedule reports
- Running times of various reports
- Inventory of tools required to produce a given part
- Operator instructions

- Program listings contained on a given disk
- Reports showing when a program was used last and how often a program is used
- Machine utilization reports
- Reports showing downtime for machines

The ability to produce management reports such as these, coupled with the other advantages of eliminating the tape medium, is making DNC the norm in NC settings.

BEST APPLICATIONS OF DNC

In spite of these advantages, DNC is not always the most appropriate NC methodology. Its best applications are in settings that require the types of management reports that a DNC system can produce and in flexible manufacturing settings. Applications that require large amounts of control information are also appropriate for DNC. These are applications that use many part programs, thus much program management is necessary.

DNC is also ideally suited for control of flexible manufacturing systems. In flexible manufacturing systems, a central host computer is needed to direct the flow of parts through the system and to download NC programs to microcomputer controllers of individual NC machines. As the technology continues to develop, DNC networks will expand to include not just machine tools, but also computer aided design and drafting systems and other computer based systems tied to production.

OVERVIEW OF CNC

The advent and full development of personal computers allowed for the development of computer numerical control (CNC). In CNC, a microcomputer controller is used to control one or more NC machines, rather than punched paper tape or magnetic plastic tape that is fed through a tape reader. The NC program is entered into the microcomputer, which, in turn, executes the program. This allows operators to develop NC programs on diskettes and load them into the microcomputer controller via a disk drive or to download programs directly from a host computer.

One advantage of CNC that will become more and more dominant as years go by and technology continues to develop is that it will allow the database created during the design and drafting of a part to be used in formulating the NC program to make the part. CNC is sometimes referred to as softwired NC. However, the term CNC is becoming the more widely used term. Another advantage of CNC is that a library of NC programs can now be created and easily stored for reuse. In addition, master copies can be made of a specific NC program so that it can be revised to create a new NC program.

ELEMENTS OF A CNC SYSTEM

A CNC system consists of three components (Figure 4–15):

1. the **control** component
2. the **input** component
3. the **output** component

The input component allows the system to receive input from an operator as well as feedback from machines interfaced as part of the system. The control component provides the same types of services that a CPU provides in any computer-controlled system. These services include memory for storage of NC programs, arithmetic and logic decisions, and communication between the input and output components. The output component carries instructions from the control component to the actual machines that will perform the work. These machines are in a closed-loop system and feed back to the controller so that their operation can be continually monitored.

As with any computer-controlled system, a CNC system may have two types of memory: **random access memory (RAM)** and **read only memory (ROM)**. NC programs are generally stored in RAM so they can be accessed easily and edited at any time. Only static programs such as the system operating program and programs used for diagnostic purposes are stored on ROM.

In a typical CNC setting, an operator enters an NC program at an operator station. This program is fed into the memory section of the controller and stored until needed. When needed, the program is called up, arithmetic and logic decisions required in the program are made, and

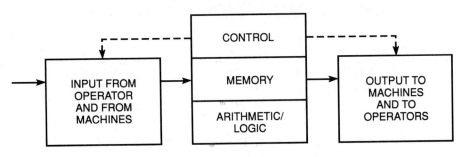

FIGURE 4–15 CNC controller interfaces.

instructions are issued by the controller to the various machines that will perform the work. In Figure 4–15, the heavy arrows moving from left to right represent the transition of data. The dotted-line arrow connecting the input and output components through the controller represents timing and control signals between the output and input components.

HISTORICAL DEVELOPMENT OF CNC

The historical development of CNC parallels the development of computers, especially personal computers. The storage capability of a personal computer was perhaps the most important development. Personal computers allow parts programs to be written using variables instead of specific values. This allows branching within programs based on the value of the variables. This allows one-part programs to be used to make a variety of parts instead of requiring a new part program for each individual part. CNC also solves the problems associated with paper and plastic tape as well as the problems associated with downtime in the host computer.

Shortly after the advent of CNC came the concept of distributed numerical control, which is now what is usually meant when the term "DNC" is used. Distributed numerical control takes advantage of the best of computer and direct numerical control. By connecting a personal computer to a host computer, manufacturing personnel have a central storage location for all programs, the ability to download from the central computer to individual microcomputers controlling one or more machines, memory in the microcomputers to keep operations in process in the event of host computer downtime, and the ability to run communications back from the individual machines to the host computer. This linkage of direct numerical control with computer numerical control to form the concept of distributed numerical control is the numerical control of the future.

CNC VERSUS DIRECT NUMERICAL CONTROL

Because CNC and direct numerical control both involve computers, they are sometimes confused. However, there are definite differences between the two with which manufacturing personnel should be familiar. In a direct numerical control system, the host computer controls many machines. However, with CNC the microcomputer controls only one or a few machines. With direct numerical control, the host computer may be in a remote location. With CNC, the microcomputer controller is at or near the workstation. Finally, the software for CNC and direct numerical control are different. Software for a direct numerical control system is usually written so that it can support a broad base of manufacturing activity, while CNC software tends to be more machine and job specific. As the concept of distributed numerical control continues to evolve, confusion between CNC and direct numerical control will be eliminated.

OPERATOR INTERFACES IN CNC

An operator interface is any device in a CNC system that is used to send or receive control information. In modern CNC systems, six types of operator interfaces can be used: paper tape reader, magnetic tape on disk or drum, punched cards, operator stations, host computer, and modem (Figure 4–16). Of these, operator stations, host computers, modems, and magnetic tape on disk or drum are becoming the most

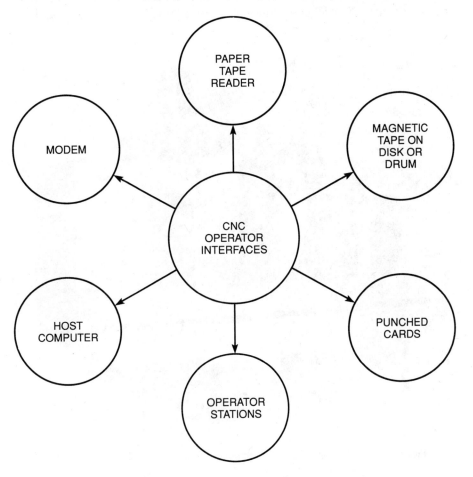

FIGURE 4–16 CNC operator interfaces.

widely used. Paper tape readers and punched cards are being seen less and less as computer technology continues to evolve. A CNC operator station contains all of the buttons, switches, and displays required to operate a CNC machine (Figure 4–17). Operator stations are used to initiate operation, to input data, and to monitor the activities of the machines. Monitoring data are displayed on a screen or through a series of status lights.

A host computer in a distributed numerical control (DNC) system is connected to individual microcomputer controllers via a data communications link. Operator data is input via magnetic tape on disk or drums or by entering it through one of the microcomputer controllers. The term **modem** is derived from the terms **modulator** and **demodulator**. A modem converts data from a CNC controller into a form that can be carried over telephone lines and vice versa. One of the principal uses of modem connections is diagnostics. For example, it is now common for manufacturers of CNC systems to provide a modem with each machine

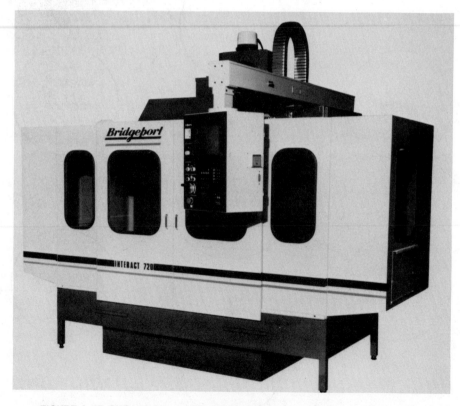

FIGURE 4–17 CNC machine with controller. *Courtesy of Bridgeport Machines, Inc.*

so that a telephone linkup can be made between the vendor and the machine as it operates on the shop floor. If a machine is not operating properly, a computer at the vendor's end of the linkup can diagnose the problem by telephone. Modem linkups for diagnostic purposes and for transmitting signals to and from remote locations will be widely used in the future.

KEY TERMS

Numerical control (NC)
Reader
Automatically programmed tools
     (APT) language
Direct numerical control (DNC)
Host computer
Programmable logic
     controllers (PLC)
Computer numerical
     control (CNC)
Data management
Distributed numerical
     control (DNC)
Machine control unit (MCU)
Communication interface
Open-loop activator
Closed-loop activators
Manual programming
Digitizing
Written programs

Graphics programs
Point-to-point machines
Continuous-path machines
Positioning machines
Contouring machines
Longitudinal axis
Transverse axis
Vertical axis
Slopes
Manual data input (MDI) control
Post-processor
Control
Input
Output
Random access memory (RAM)
Read only memory (ROM)
Modem
Modulator
Demodulator

Chapter Four REVIEW

1. What are the classifications of NC machines?
2. List four advantages usually gained from a conversion to NC.
3. List four disadvantages associated with a conversion to NC.
4. Explain the term MDI Control.
5. Explain the difference between open-loop and closed-loop activators.
6. Explain the difference between point-to-point and continuous-path machines.

7. What are the letter designations of the following axes:
   Longitudinal
   Transverse
   Vertical
8. What is a post-processor and what purpose does it serve in an NC system?
9. Explain the problem of data management in NC.
10. What is the APT language?
11. What are the basic components of a typical NC system?
12. Explain the term slopes.
13. Explain the difference between direct numerical control and distributed numerical control.
14. What is the principal shortcoming of direct numerical control?
15. What is a graphic NC program?

These case studies were provided by the Society of Manufacturing Engineers (SME). They are excerpted from the SME's *Manufacturing Insights*® series of videotapes. These case studies contained herein give students actual examples of the way the advanced manufacturing technologies covered in this chapter are being applied in "real-world" manufacturing settings.

As you read each case study, relate the examples cited to the material presented in the text of the chapter. This combination of textbook information and real-world examples will be particularly valuable to your understanding of advanced manufacturing technologies.

LOCKHEED–GEORGIA

A flexible manufacturing cell (FMC), in daily operation at Lockheed–Georgia, is programmed by a regular production numerical control (NC) programmer and run by specially trained production operators. The two main components of the cell are a robot with a multifunctional end-of-arm tooling and a workholding table, which is part of the workstation shuttle. The robot's unique end-of-arm tooling includes an airpower drill, router, and saw.

The robot's work area is protected by a fence, and the gates are electronically interlocked with the robot's hold signal. The operator selects the correct pair of templates or router blocks for the new work order and positions them in the workstation. After the corresponding program data is loaded into the robot controller's memory, the operator positions a stretch-formed extrusion on the router blocks and clamps it into place.

When ready, the operator initializes the robot to begin machining the part. The robot turns its end-of-arm tooling to the router and begins routing the profile of the part. The robot also cuts the lightening holes with its router. When it has completed this task, it turns its saw to the precise angle and cuts off each end of the extrusion. The workstation shuttles are cycled, bringing the completed part back to the load position for removal. The next extrusion is then sent forward to the robot, ready for machining, as the process continues.

A record 98.4% uptime was established during the first month of production of this Lockheed–Georgia cell. Research indicates the cell has produced other significant benefits, including a 75% savings on total production time, which includes more consistent quality, increased safety, and reduced scrap rework time.

(continued)

Lockheed–Georgia is well satisfied with this FMC, whose success has been attributed to the combination of new, yet proven, technology with an appropriate application for that technology.

BADGER METER

A flexible manufacturing cell (FMC) at Badger Meter, designed by Kearney and Trecker, was installed with plant modernization as the major business emphasis. It is a flexible, palletized cell, permitting the production of a family of components, in random order, on a to-order basis—ideal for Badger Meter's requirements.

The cell links two machining centers with a linear transport module and cell controller. It also includes a 120-tool change magazine, with broken tool identification system. The shuttle car has two arms to handle delivery to the machining centers, as well as 30 cue stations on both sides of the track. Cell control is handled by a Microvax computer, for which there are two data entry stations. Jobs can be entered at Production Control, where the operator schedules each job, including part number, part quantity, and start and finish dates. This information is then downloaded to the cell controller. The computer prioritizes the jobs on the system, assigns the route batch, and then, depending on the part sequence, the shuttle car pulls the fixtures, delivers to load or unload, or takes them to the correct station for machining.

The former method involved 13 people, while the current cell requires just 3½ people. Even though the cell processes some 75 parts, all setup has been virtually eliminated, and the average cycle time has been reduced by 33%.

In terms of total time making parts versus machine availability, the numerically controlled cell has improved utilization from 55 to 85%. Thus, the FMC has lived up to its promise at Badger Meter.

Major Topics Covered

- Robot Defined
- Robot Terminology
- Historical Development of Robots
- The Robot System
- Categories of Robots
- Selection of Robots
- Robot End-Effectors
- Robot Sensors and Machine Vision Systems
- Robot Programming and Control
- Safety and Robots
- *Case Study: Comdial*
- *Case Study: Ford Motor Company*
- *Case Study: Delta Faucet*
- *Case Study: Northern Telecom*
- *Case Study: IBM*

Chapter Five

Automated Assembly

Automated Guided Vehicles

CAD/CAM

Numerical Control

Industrial Robots

Lasers in Manufacturing

Programmable Logic Controllers

Flexible Manufacturing

Computer Integrated Manufacturing

Other Related Technologies

ROBOT DEFINED

There are a variety of definitions of the term **robot**. Depending on the definition used, the number of robot installations worldwide varies widely. Numerous single-purpose machines are used in manufacturing plants that might appear to be robots. These machines are hardwired to perform a single function and cannot be reprogrammed to perform a different function. Such single-purpose machines do not fit the definition for industrial robots that is becoming widely accepted. This definition was developed by the Robot Institute of America:

> A robot is a reprogrammable multifunctional manipulator designed to move material, parts, tools, or specialized devices through variable programmed motions for the performance of a variety of tasks.

Note that this definition contains the words **reprogrammable** and **multifunctional**. It is these two characteristics that separate the true industrial robot from the various single-purpose machines used in modern manufacturing firms. The term "reprogrammable" implies two things: The robot operates according to a written program, and this program can be rewritten to accommodate a variety of manufacturing tasks.

The term "multifunctional" means that the robot can, through reprogramming and the use of different end-effectors, perform a number of different manufacturing tasks. Definitions written around these two critical characteristics are becoming the accepted definitions among manufacturing professionals. Figures 5–1 and 5–2 are examples of modern industrial robots that fit this description.

Robots are not the highly intelligent mechanical humans many laypeople envision. They are computer-controlled manufacturing machines that can be tooled up and programmed to perform a number of different tasks.

Robots are available in four broad geometric configurations (Figure 5–3):

1. cartesian coordinates
2. cylindrical coordinates
3. polar coordinates
4. revolute coordinates

Figure 5–4 illustrates the wrist movements or articulations of an industrial robot. All four of the configurations shown in Figure 5–3 provide for movement in three axes. The three additional articulations at the wrist shown in Figure 5–4 give the modern industrial robot the potential for six separate articulations for locating parts and positioning tools.

FIGURE 5–1 Industrial robot. *Courtesy of Unimation, Inc.*

FIGURE 5–2 Industrial robot. *Courtesy of Cincinnati Milacron.* Note: Safety equipment may have been removed or opened to clearly illustrate the product and must be in place prior to operation.

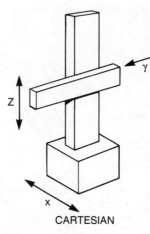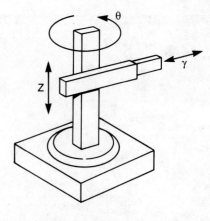

CARTESIAN CYLINDRICAL

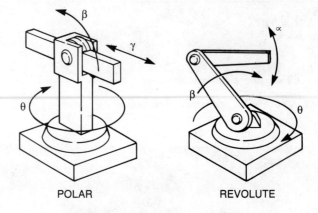

POLAR REVOLUTE

FIGURE 5-3 Robot arm configurations.

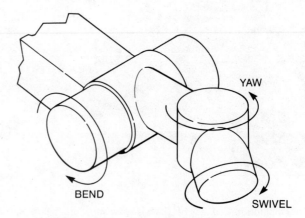

YAW

BEND

SWIVEL

FIGURE 5-4 Wrist movements of an industrial robot.

ROBOT TERMINOLOGY

Several terms are used frequently in any discussion of industrial robots. The following definitions should be studied before proceeding with the remainder of this chapter.

Accuracy: A measure of how close a robot arm is to come to the coordinates specified. There is always some difference between the actual and the desired point. The degree of difference is the accuracy of the robot.

Accuator: Any device in a robot system that converts electrical, hydraulic, or pneumatic energy into mechanical energy or motion.

Continuous Path: A servo-driven robot that provides absolute control along an entire path of arm motion, but with certain restrictions on the degree of difficulty in changing the program.

Controlled Path: A servo-driven robot with a control system that specifies the location and orientation of all robot axes. A control-path robot moves in a straight line between programmed points.

Degrees of Freedom: The number of degrees of freedom of a robot is the number of movable axes on the robot's arm. A robot with four movable joints has four degrees of freedom.

End-Effector: An end-of-arm tool that is attached to the robot's manipulator and actually performs the robot's work.

Limited Sequence: A simple, non-servo type of robot, sometimes called a bang-bang robot. Movement of a limited-sequence robot is controlled by a series of stop switches.

Manipulator: Another name for the arm of the robot. It encompasses basic axes that control wrist movements for robots. The three basic axes are referred to as the pitch, yaw, and roll.

Payload: The maximum weight a robot is able to carry at normal speeds.

Pitch: Up-and-down motion along an axis.

Point-to-Point: A robot with a control system for programming a series of points without regard to coordination of axes.

Repeatability: The degree to which a robot is able to return the tool center point repeatedly to the same position.

Roll: Circular motion along an axis.

Servo-Mechanism: An automatic feedback control system for mechanical motion.

Teach Pendant: A special control box that an operator uses to guide a robot through the motions required to perform a specific task.

Tool Center Point: A given point at the tool level around which the robot is programmed to perform specific tasks.

Work Envelope: The operating range or reach capability of a robot.

Yaw: Side-to-side motion along an axis.

HISTORICAL DEVELOPMENT OF ROBOTS

The history of robots goes back to the first attempts to automate manufacturing. Although robots that fit the definition presented have been in existence about 20 years, students of Advanced Manufacturing Technology should be familiar with the historical developments that led to the current state of the art in industrial robots.

The first articulated arm came about in 1951 and was used by the U.S. Atomic Energy Commission. In 1954, the first programmable robot was designed by George Devol. It was based on two important technologies:

1. numerical control (NC) technology
2. remote manipulation technology

Devol coined the buzzword universal automation, which was shortened to unimation and became the name of the first commercial industrial robot vendor, a company founded by Joseph Engelberger. Figures 5–5 and 5–6 are examples of modern industrial robots produced by Unimation.

Numerical control technology provided a form of machine control ideally suited to robots. It allowed for the control of motion by stored programs. These programs contain data points to which the robot sequentially moves, timing signals to initiate action and to stop movement, and logic statements to allow for decision making.

Remote manipulator technology allowed a machine to be more than just another NC machine. It allowed such machines to become robots that can perform a variety of manufacturing tasks in both inaccessible and unsafe environments. By merging these two technologies, Devol

FIGURE 5-5 Industrial robot. *Photo by David L. Goetsch.*

developed the first industrial robot, an unsophisticated programmable materials handling machine.

The first commercially produced robot was developed in 1959. In 1962, the first industrial robot to be used on a production line was installed by General Motors Corporation. This robot was produced by Unimation. A major step forward in robot control occurred in 1973 with the development of the T-3 industrial robot by Cincinnati Milacron. The T-3 robot was the first commercially produced industrial robot controlled by a minicomputer. Cincinnati Milacron is still a leading vendor of industrial robots.

Since the late 1960s and early 1970s, the number of robot installations in manufacturing firms has grown rapidly. Figure 5-7 is a chart that indicates the current and projected growth of industrial robot installations in the United States through the year 2000. The data were collected and projected by the industrial Robot Division of Cincinnati Milacron.

Note that in 1965 there were well under 5,000 industrial robot installations. By 1985, this number had grown to 15,000. But the most rapid growth is projected to take place between 1985 and 2000, when there will be over 100,000 industrial robots installed in manufacturing plants in this country at the turn of the century.

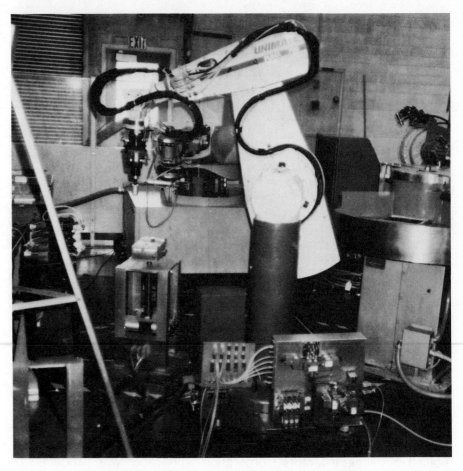

FIGURE 5–6 Industrial robot. *Photo by David L. Goetsch.*

RATIONALE FOR INDUSTRIAL ROBOTS

Numerical control and remote manipulator technology prompted the wide-scale development and use of industrial robots. But major technological developments do not take place simply because of such new capabilities. Something must provide the impetus for taking advantage of these capabilities. In the case of industrial robots, the impetus was economics.

The rapid inflation of wages experienced in the 1970s tremendously increased the personnel costs of manufacturing firms. At the same time, foreign competition became a serious problem for U.S. manufacturers. Foreign manufacturers who had undertaken automation on a wide-scale basis, such as those in Japan, began to gain an increasingly large share of the U.S. and world market for manufactured goods, particularly automobiles.

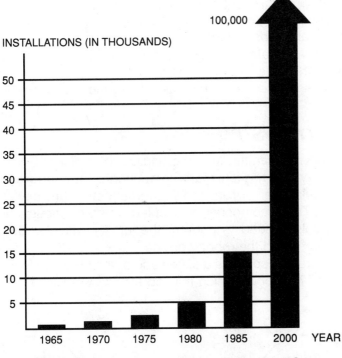

INSTALLATIONS (IN THOUSANDS)

100,000

FIGURE 5-7 Growth of robot installations in the United States.

Through a variety of automation techniques, including robots, Japanese manufacturers, beginning in the 1970s, were able to produce better automobiles more cheaply than nonautomated U.S. manufacturers, who were continually engaged in struggles with organized labor. Consequently, in order to survive, U.S. manufacturers were forced to consider any technological developments that could help improve productivity.

It became imperative to produce better products at lower costs in order to be competitive with foreign manufacturers. Other factors such as the need to find better ways of performing dangerous manufacturing tasks contributed to the development of industrial robots. However, the principal rationale has always been, and is still, improved productivity.

Industrial robots offer a number of benefits that are the reasons for the rapid current and projected growth of industrial robot installations:

1. increased productivity
2. improved product quality
3. more consistent product quality
4. reduced scrap and waste

5. reduced reworking costs
6. reduced raw goods and inventory
7. direct labor cost savings
8. savings in related costs such as lighting, heating, and cooling
9. savings in safety-related costs
10. savings from correctly forecasting production schedules

THE ROBOT SYSTEM

Work in a manufacturing setting is accomplished by a robot system. A robot system has four major components: the controller, the robot arm or manipulator, end-of arm tools or end-effectors, and power sources (Figure 5–8). These components, coupled with the various other equipment and tools needed to perform the job for which a robot is programmed, are called the robot **work cell**.

Figure 5–9 contains a schematic of the work cell for Cincinnati Milacron's T3-726 robot, which is used for TIG welding. Note that the

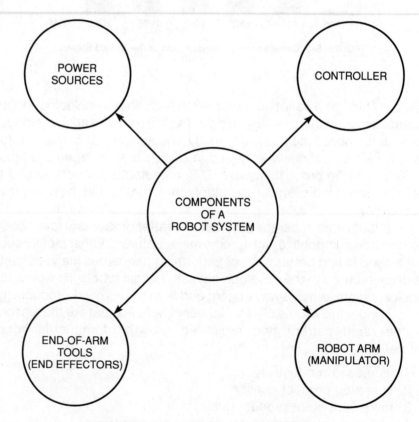

FIGURE 5–8 Components of a robot system.

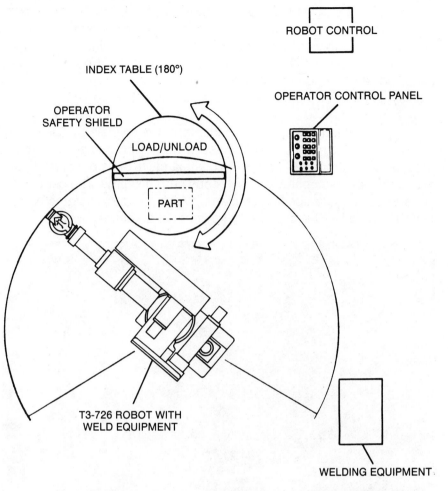

ROBOT CONTROL

INDEX TABLE (180°)

OPERATOR CONTROL PANEL

OPERATOR
SAFETY SHIELD

LOAD/UNLOAD

PART

T3-726 ROBOT WITH
WELD EQUIPMENT

WELDING EQUIPMENT

FIGURE 5–9 Robot work cell. *Courtesy of Cincinnati Milacron—Industrial Robot Division.*

work cell contains not just the robot system, but also an index table with an operator safety shield and special welding equipment.

The contents of a robot work cell varies according to the application of the robot. However, the one constant in a robot work cell is the robot system, made up of a controller, robot arm, end-of-arm tools, and power sources.

The Controller

A robot controller is a special-purpose device similar to a computer. As such, it has all of the usual components of a computer, including a central processing unit made up of a control section and an arithmetic/logic section, and a variety of input and output devices.

The controller for a robot system does not look like the micro-computer one is used to seeing on a desk. It must be packaged differently to be able to withstand the rigors of a manufacturing environment. This is known as being **factory hardened**. Typical input/output devices used with a robot controller include teach stations, teach pendants, a display terminal, a controller front panel, and a permanent storage device.

Teach terminals, teach pendants, or front panels are used for interacting with the robot system. These devices allow humans to turn the robot on, write programs, and key in commands to the robot system. Display terminals give operators a soft copy output source. The permanent storage device is a special device on which usable programs can be stored. Figure 5–10 shows a robot system and controller.

FIGURE 5–10 Industrial robot with controller. *Photo by David L. Goetsch.*

Mechanical Arm

The mechanical arm, or manipulator, is the part of the robot system with which most people are familiar. It is the part that actually performs the principal movements in doing manufacturing work. Mechanical arms are classified according to the types of motions they can do. The basic categories of motion for mechanical arms are rectangular, cylindrical, and spherical. Figure 5–11 shows the mechanical arm for the Unimate Puma 762 robot.

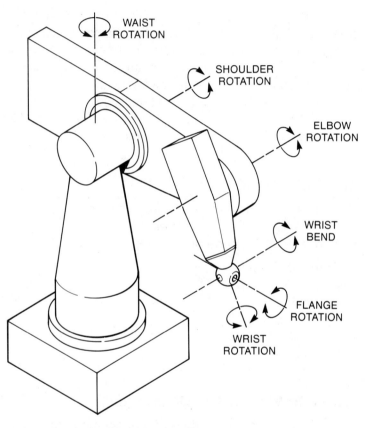

FIGURE 5–11 Mechanical arm.

End-Effectors

The human arm by itself can perform little work. It needs a hand to grip tools and materials. The mechanical arm of a robot also needs a hand and a tool, or a device that contains both functions. On industrial robots, the function of the wrist is performed by the tool plate, a special device to which end-effectors are attached. The tools themselves vary according to the type of tasks the robot will perform. Any special device or tool attached to the tool plate to allow the robot to perform some specialized task is classified as an end-effector. End-effectors include grippers, holders, and various other specialized tools. Most end-effectors are designed with a special purpose in mind. However, a well-designed end-effector is versatile enough to handle a variety of different materials and more than one task.

The Power Source

A robot system can be powered by one of three different types of power: electrical, pneumatic, or hydraulic. The controller, of course, is powered by electricity, as is any computer. The mechanical arm and end-effectors may be powered by either pneumatic or hydraulic power. Some robot systems use all three types of power. For example, a robot might use electricity to power the controller, hydraulic power to manipulate the arm, and pneumatic power to manipulate the end-effector. Hydraulic power is fluid based; pneumatic power comes from compressed gas.

CATEGORIES OF ROBOTS

There are no industry standards that mandate a given approach to categorizing robots. However, it has become standard practice to categorize robots in one of six ways:

1. arm geometry
2. power sources
3. applications
4. control technique
5. path control
6. intelligence

Categorizing Robots by Arm Geometry

There are four classifications of arm geometry for industrial robots: rectangular, cylindrical, spherical, and jointed arm. A robot with rectangular arm geometry has a rectangular work envelope if viewed orthographically. If viewed in three dimensions, the work envelope is shaped like an elongated cube. An industrial robot with cylindrical arm geometry has a work envelope described by a cylinder. An industrial robot with spherical arm geometry has a work envelope described by a portion of a sphere. The movement of a jointed arm robot is similar to the movement of a human arm. The joints mimic such human joints as the shoulder, elbow, and wrist. The work envelope is a quasi-sphere.

Each type of coordinate system has its individual characteristics as well as its advantages and disadvantages. A better understanding of industrial robots can be gained by closely analyzing the coordinate systems on which their work envelopes are based.

There are three types of coordinate systems associated with industrial robots: rectangular, cylindrical, and spherical. As the names imply, these systems have rectangular, cylindrical, and spherical work envelopes, respectively.

A rectangular work envelope is defined by imaginary lines running parallel to the X, Y, and Z axes. Advantages of this type of geometry include (1) the potential for large work envelopes, (2) overhead mounting that frees up space on the shop floor, and (3) the allowance for simple, less sophisticated controllers. However, overhead mounting structures, when used, can limit the access of materials handling equipment, and maintenance of overhead drive and control equipment is difficult because access is limited.

A cylindrical work envelope is defined by imaginary lines in the form of a vertical cylinder. Advantages of this type of robot include (1) deep horizontal reach into bins or machines, (2) it requires relatively small amount of floor space, and (3) it has the potential to handle heavy payloads. Cylindrical robots do have the disadvantage of limited left and right reach.

The work envelope of spherical coordinate robots may have a variety of shapes. Spherical geometry robots have the same advantages and disadvantages as cylindrical robots.

Categorizing Robots by Power Source

There are three power sources for industrial robots: electric, hydraulic, and pneumatic. All robots require electric power to operate the controller. The manipulator and end-effector may be operated by electric, pneumatic, or hydraulic power, or a combination of these. A common configuration is one in which electric power is used to operate the manipulator and pneumatic power is used to operate end-of-arm tools; but this varies from robot to robot. Servo-electric drives are the most widely used for assembly applications, a rapidly growing application.

Electric Power

Electric power is used to power the controllers in all industrial robot systems. However, electricity can also be used to power the manipulator and end-effector in applications with small payloads. Electrically powered robots use an electric motor servo-circuit to drive the manipulator. This servo-circuit uses a potentiometer, transducer, optical device, or a resolver to determine and continually communicate the position of the tool. Electric robots are able to perform tasks that require a high degree of accuracy and repeatability. With electrical power, there is a comparatively low noise level, no problems with oil leaks and spills, and the ability to adapt to sophisticated controls; and they are relatively inexpensive. Electrically driven robots are normally confined to applications involving payloads of 200 pounds or less and they are sometimes prone to overshooting.

Hydraulic Power

Hydraulic power is provided by pressurized oil or some other type of fluid. Hydraulically powered robots have a number of disadvantages: They develop oil leaks, which sometimes cause oil spills in the work area; they are loud; and they may pose a fire hazard due to the oil leaks. However, in spite of these disadvantages, hydraulic power can achieve a very high power-to-size ratio. This means that hydraulically powered robots can handle much larger payloads than electrically powered robots. For this reason, the hydraulically powered robot is the most frequently used in heavy-payload applications. In addition, hydraulically powered robots are more resilient than electrically powered robots. They have the ability to give when they make hard contact with an immovable surface.

Pneumatic Power

Pneumatic power is provided by pressurized gas. The configuration of a pneumatically powered robot is similar to that of a hydraulically powered robot, except that pressurized gas is used instead of pressurized oil. This solves the problem of oil leaks and oil spills in the working area. Also, they are relatively inexpensive. However, at present, the feedback control systems for pneumatically powered robots are not well developed. Consequently, pneumatically powered robots tend to be those that are referred to as bang-bang robots. Pneumatic power is much cleaner than hydraulic power, and it is better in high-velocity applications where environmental considerations rule out the use of hydraulic power.

Categorizing Robots by Application

Another way to categorize robots is by application. All of the various robot applications can be classified as either assembly or nonassembly applications. Applications such as welding, palletizing, stacking, unstacking, loading, materials handling, drilling, grinding, deburring, painting, and gluing are all nonassembly applications.

Assembly applications involve such tasks as soldering, press fitting, welding, and applying threaded fasteners. Assembly applications typically require high accuracy and repeatability, but have low payload capacities. For this reason, most robots used in assembly applications are electrically powered. Nonassembly robots are typically hydraulically or pneumatically powered because of the large payloads involved.

Categorizing Robots by Control Systems

When categorizing robots according to control systems, there are two broad categories: servo and non-servo control robots. Servo-controlled robots are also referred to as closed-loop systems. Servo system robots are more sophisticated and, in turn, more expensive. Non-servo system robots are simple and relatively inexpensive. In both cases, the control system refers to the method used to control the position of the robot tool.

Servo, or closed-loop, systems use a sensor device to continuously monitor the position, direction, and velocity of the robot tool as necessary to keep the tool on the desired path. Servo systems are used in those applications where path control is a critical element. Because of the feedback capability, servo or closed-loop robots can perform more complex manufacturing tasks and a wider variety of tasks.

Non-servo, or open-loop, robots do not employ sensors to monitor the position, direction, and velocity of the robot tool. Rather, a series of physical limits or fixed stop points are used. Each joint in the robot must have these physical limits or fixed stops built in. The robot motion is from one extreme, or stop point, to the next extreme, or stop point. There are no intermediate stops in a non-servo robot. Non-servo robots are also referred to as pick-and-place or bang-bang robots.

These robots require less maintenance and are less expensive. However, they are limited to the simpler manufacturing tasks, require more complex end-effectors, and tend to have a shorter life span than servo-controlled robots.

Categorizing Robots by the Type of Path Control

Path control refers to the method used by the robot controller to guide the end-effector along the desired path. There are four steps of path control used in modern industrial robots: stop to stop, continuous path, point to point, and controlled path.

Stop-to-stop path control is used in non-servo control robots. This type of robot controls the path of the tool with a series of preset electronic or mechanical stop switches, which establish the extremes of motion for the robot and are responsible for the robot's ability to work its way through a sequence of motions required to complete a manufacturing task. Because of this simple, unsophisticated method of path control, stop-to-stop robots are generally used to perform only simple materials handling and line transfer tasks.

Point-to-point, continuous-path, and control-path systems are used with servo-controlled robots. Point-to-point path control is accomplished by programming a series of positions along a desired path of motion. Only stop points are specified. Point-to-point path control robots are used for materials handling and other applications where the control of movement between points is not a critical factor.

Continuous-path control robots are used in applications where the path of the tool is critical. With this type of control, every point along the path of the tool is critical. For this reason, the controller for a continuous-path robot requires more memory than a point-to-point control robot. Every point along the desired path must be stored in the memory of the controller. Continuous-path control is generally not used in applications that involve frequent changes to the program.

Controlled-path robots are those in which all axes along a path are coordinated. In this method, the path and the velocity at which the tool moves along the path are controlled. Controlled-path robots are typically used in processing, assembly, soldering, and welding applications.

Categorizing Robots by Levels of Intelligence

When robots are categorized according to their level of intelligence, there are three categories: high-technology, medium-technology, and low-technology robots. Pick-and-place or bang-bang robots are considered low-technology robots. High-technology robots are servo-controlled robots that use either continuous-path or controlled-path control systems. It is more difficult to categorize robots in the intermediate group. These tend to be the more flexible non-servo robots and the simpler point-to-point servo robots.

ROBOT END-EFFECTORS

End-effectors are the end-of-arm tools that actually do the work of the robot. The most frequently used end-effectors come in two broad categories: grippers and special-purpose end-effectors. Grippers, as the name implies, are used for gripping an object, performing the desired, movement, and releasing the object. There are five different classifications of grippers: standard, vacuum, magnetic, air-pressure, and special-purpose (Figure 5–12).

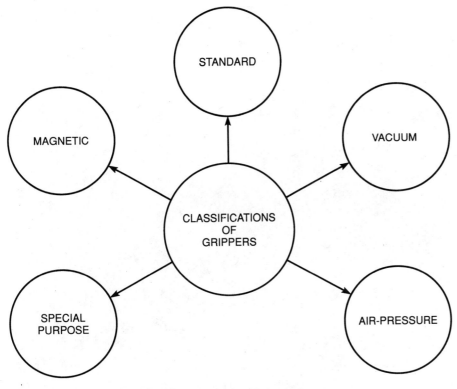

FIGURE 5–12 Classifications of grippers.

Standard grippers grasp the object between two fingers, somewhat like a mechanical claw. Figure 5–13 is an example of an industrial robot with standard grippers as end-effectors. Vacuum end effectors use vacuum suction to pick up materials and magnetic grippers use magnetism. Air-pressure grippers use pneumatic fingers, which are particularly effective in applications requiring a gentle touch. Special-purpose grippers are designed for applications where standard, vacuum, magnetic, and air-pressure grippers are not appropriate. Many special-purpose grippers are designed locally by the robot user.

Special-purpose end-effectors are not grippers; they are special-purpose tools designed to do specific jobs, such as drilling, welding, painting, grinding, and sanding. Figure 5–14 is an example of a robot with a special-purpose end-effector for welding.

In addition to grippers, special tools can also serve as end-effectors. Such tools are usually fastened to the robot wrist and include such special-purpose tools as the following:

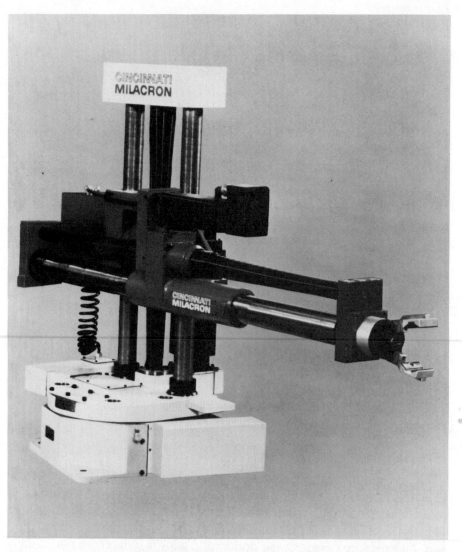

FIGURE 5–13 Industrial robot with standard gripper. *Courtesy of Cincinnati Milacron.* Note: Safety equipment may have been removed or opened to clearly illustrate the product and must be in place prior to operation.

- Welding tools (spot and arc)
- Spray painting tools
- Drilling tools
- Routing tools
- Grinding tools

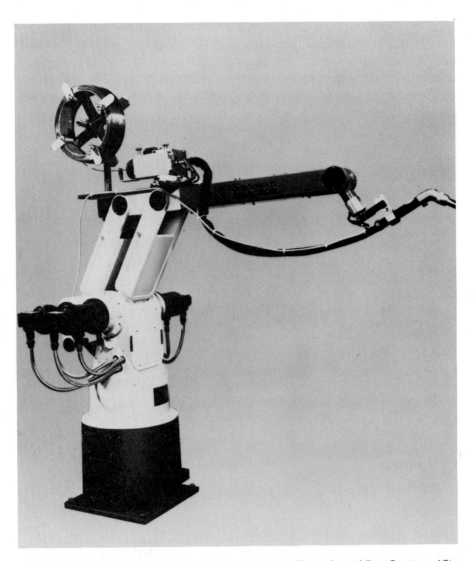

FIGURE 5–14 Industrial robot with a special-purpose end effector for welding. *Courtesy of Cincinnati Milacron.* Note: Safety equipment may have been removed or opened to clearly illustrate the product and must be in place prior to operation.

ROBOT SENSORS AND MACHINE VISION SYSTEMS

People are aware of their surroundings because they have five senses: sight, hearing, smell, taste, and touch. Using these senses, manufacturing personnel are able to perform the tasks required of them. Even the simplest task, such as picking up a tool, requires the senses of sight and touch.

Like people, robots must be aware of their environment if they are to perform manufacturing tasks. Sensors are devices that make robots aware. There are three basic types of robot sensors: contact, noncontact, and process-monitoring sensors (Figure 5–15). Before examining each type, the reader should become familiar with the purposes of robot sensors.

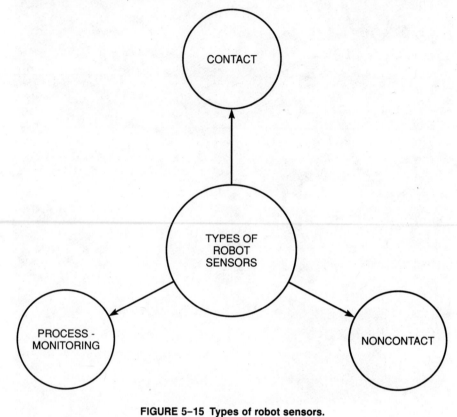

FIGURE 5–15 Types of robot sensors.

Purposes of Robot Sensors

Robot sensors serve a variety of purposes, including locating and identifying parts to be assembled, verification of the presence and the proper orientation of parts, determining forces, measuring temperatures, correction of errors in positioning, regulation of gripper pressure, protection against overloads, and calibration/recalibration of robots.

All of these purposes fall into one of four broad categories:

1. monitoring
2. detection
3. analysis
4. calibration

In the first category, sensors allow robots to monitor the quality, consistency, and conformity of parts. They also allow robots to respond to factors that are important in assembly operations, such as identifying the proper parts, ensuring that parts are properly located in the assembly, and ensuring that parts are properly oriented before putting them in the assembly. Parts location, identification, and orientation fall into this category.

In the second category, sensors allow robots to detect circumstances that could be dangerous to the robot, the product, other equipment in the work cell, or to human workers. They also allow robots to detect the presence of a part, as well as system breakdowns or malfunctions.

In the third category, sensors allow robots to analyze parts for quality considerations. They also allow robots to analyze the robot system for internal problems so that the problem can be quickly and easily corrected. In the final category, sensors can provide the positioning information needed to recalibrate a robot for a greater degree of accuracy.

Contact Sensors

Contact sensors are those that require the end-effector to actually touch the part. The most frequently used type of contact sensor is the limit switch. This is the least sophisticated type of sensor.

Limit switches are open-loop, non-servo devices used primarily with bang-bang or pick-and-place robots to detect motion and movement of parts within the work cell. There is no feedback channel between the end-effector and the controller. Limit switches, like light switches, are electrical devices that are turned on and off mechanically by an actuator. The actuator is usually a part called a **dog**.

A more sophisticated type of contact switch is artificial skin. Artificial skin is a closed-loop, servo-controlled gripper lining. A robot with artificial skin can vary its gripper pressure according to the physical properties of the part. This capability is particularly important for robots used in assembly operations.

Artificial skin is referred to as **tactile sensing**. Tactile sensing involves using a group of sensors arranged in a rectangular matrix or array. Each element in the matrix can sense a part and send feedback to the controller about such factors as the part shape, texture, position, and orientation.

Noncontact Sensors

Contact sensors, as you have seen, are similar to the human sense of touch in that contact must be made with the part in order for sensing to take place. Noncontact sensors, on the other hand, relate more closely to such senses as sight, hearing, or smell in that they do not require contact between the part and the robot. There are three principal types of noncontact sensors: proximity, photo-optic, and vision (Figure 5–16).

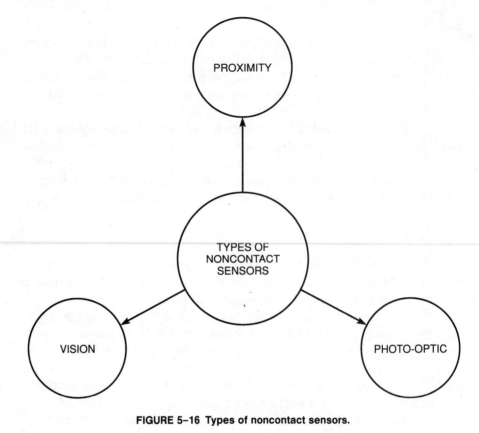

FIGURE 5–16 Types of noncontact sensors.

Proximity Sensors

Proximity sensors detect the presence of a part in a specified electromagnetic or electrostatic field. The actual range of the sensor field varies from model to model and type to type. Proximity sensors come in a number of shapes. The shape that works best in a given application depends on the shape of the part and the individual needs of the specific work cell.

Proximity sensors are available for sensing both metallic and nonmetallic parts. Metal-sensing proximity devices use a high-frequency electromagnetic field to sense the part. Nonmetal-sensing proximity devices use electrostatic capacitance as the sensing medium.

Photo-Optic Sensors

Photo-optic sensors detect a part when it reflects back or breaks a light beam. This technology is similar to that used in some burglar alarm systems. There are four principal types of photo-optic sensors: separate, retro-reflective, diffuse-reflective, and definite-reflective sensors.

Separate sensor devices take their name from their configuration. This type of sensor system consists of two separate devices, one to produce and direct the light beam and one to receive it. When the receiver fails to detect the presence of the light beam, it knows the beam has been broken.

Retro-reflective sensors are similar to separate sensors except they use two light beams, one outgoing and one returning beam. Such a system has two hardware components: (a) a device to produce and send the outgoing light beam as well as to receive the beam that is reflected back, and (b) a device to receive the outgoing light beam and reflect it back to the source. Sensing occurs when a part breaks either the outgoing or the reflected light beam.

Diffuse-reflective sensors bounce a light beam off the actual part. Such a system has only one hardware component, a special device that produces and directs the light beam, then receives diffuse light reflected back from the part. A reflected beam of light is not necessary; diffuse reflected light is sufficient for detection to occur.

Definite-reflective sensors are similar to both diffuse-reflective and retro-reflective sensors. They are like diffuse-reflective sensors in that they use a single hardware device with both sending and receiving capabilities and they bounce the light off the actual object. They are like retro-reflective sensors in that what is reflected back must be a bona fide light beam; diffuse light is not sufficient to allow for detection.

Of the four types of photo-optic sensors, separate sensors have the longest range. In descending order of range, the other sensors are retro-reflective, diffuse-reflective, and definite-reflective.

Vision Sensors

Vision sensors are the newest and most sophisticated of the noncontact sensors. Vision sensor systems consist of three components:

1. a camera and controller
2. interface circuitry
3. display terminal

The camera directed by the controller sees the silhouette of the part. The profile of the part is then displayed on the screen of the display terminal. The interface circuitry joins the camera and the controller to the work cell and other systems interfaced with the work cell.

Vision systems allow robots to see what takes place in the workcell. Typical uses of vision systems include part orientation, inspection, location, and identification. Robot vision has enormous potential for expanding the capabilities and applications of industrial robots. However, it is a relatively new technology and much research is still needed. The current state of the art is weak in such areas as recognition of three-dimensional images, selection of parts from nonordered bins, and selection of random parts from within the same bins.

Process Monitoring Sensors

All of the sensors discussed so far have been robot sensors, which make a robot a more intelligent component of the work cell. Process monitoring sensors are not robot sensors. Rather, they are used for monitoring other manufacturing operations that interface with the work cell. When a problem arises in a manufacturing operation that is interfaced with the robot cell, process monitoring sensors warn the robot that corrective action is required.

ROBOT PROGRAMMING AND CONTROL

Like computers, robots are capable of doing only what they are told. Also like computers, the communication link between robots and people is the program. In other words, people tell robots what to do through programs.

Robot programming languages have not yet been developed to the extent of computer languages. There are no standards for robot programming languages at present. Consequently, each vendor develops its own language for its own robot. Standardization of robot programming languages is a high-priority goal of the robot industry.

Joint Control Languages

Joint control languages are the lowest level of robot programming languages. They control robot motion by controlling the positions of the manipulator in all axes for all joints. Joint positions are specified in angular form. Joint control languages are used with only the least sophisticated robots. A major weakness of joint control languages is that they are written in terms of the motion of the robot manipulator at its various joints rather than in terms of solving the manufacturing problem.

Primitive Motion Languages

Primitive motion languages are the next higher level and the most widely used of the robot programming languages. Primitive motion languages are also called point-to-point languages. To write a point-to-point program, the robot manipulator must actually be moved to the various points that define the motion sequence. Each time a desired point is reached, a switch is activated, making that point part of the program.

Although primitive motion languages are widely used, they have a number of disadvantages. The most important of these are that they cannot be written off-line and they focus on the motion of the robot instead of on the task to be performed.

Structured Languages

Structured languages are the next highest level category of robot programming languages. The various languages in this category begin to resemble computer programming languages, primarily due to their ability to use coordinate transformations and because they allow for off-line programming.

Like computer programming languages, structured robot programming languages allow for branching and subroutines. They also resemble computer programming languages in that they require specialized training. This factor must be considered a negative. However, when weighed against the many advantages gained from structured languages—the use of transformations, the off-line programming capability, and branching and subroutine capabilities—the training factor becomes less important.

Task-Oriented Languages

Task-oriented languages are the highest level of robot programming languages. The languages allow the user to actually program the robot in terms of the manufacturing problem rather than the motion of the manipulator. Programmers are able to use such normal conversational terms as "insert bolt" and "tighten nut". Therefore, task-oriented languages are easy and convenient for people to learn.

However, task-oriented languages are not yet well developed, nor are they widely available at present. Much research remains to be done at this level of robot programming. However, several task-oriented languages are currently in various stages of development. Robot productivity will reach new heights when task-oriented programming languages are fully developed and utilized.

The actual methods used for programming robots are the manual method, the walk-through method, the lead-through method, and the off-line programming. Of these, the most efficient is the off-line programming for computer numerical control. This is done using the types of languages described in the previous paragraphs. With off-line programming, a program for guiding the robot through the various required motions is written using a personal computer and stored. With off-line programming, unlike the other three methods, the robot does not have to be stopped from its work in order to be programmed. With off-line programming, the program can be written while the robot is working on another operation.

SAFETY AND ROBOTS

One of the principal advantages of robots is that they can be used in settings that are dangerous to humans. Welding and parting are examples of applications where robots can be used more safely than humans. Even though robots are closely associated with safety in the workplace, they can, in themselves, be dangerous.

Robots and robot cells must be carefully designed and configured so that they do not endanger human workers and other machines. Robot work envelopes should be accurately calculated and a danger zone surrounding the envelope clearly marked off. Red flooring strips and barriers can be used to keep human workers out of a robot's work envelope.

Even with such precautions, it is still a good idea to have an automatic shutdown system in situations where robots are used. Such a system should have the capacity to sense the need for an automatic shutdown of operations. Fault-tolerant computers and redundant systems can be installed to ensure proper shutdown of robotics systems to ensure a safe environment.

KEY TERMS

Robot
Reprogrammable
Multifunctional
Accuracy
Actuator
Continuous path
Controlled path
Degrees of freedom

End-effector
Limited sequence
Manipulator
Payload
Pitch
Point-to-point
Repeatability
Roll

Servo-mechanism
Teach pendant
Tool center point
Work envelope
Yaw

Work cell
Factory hardened
Dog
Tactile sensing

Chapter Five REVIEW

1. Define the term industrial robot.
2. List five benefits of industrial robots.
3. List and explain the four main components of a robot system.
4. Explain how robots can be classified according to arm geometry.
5. Explain the three power sources available to robots.
6. Explain servo and non-servo control systems and how they differ.
7. Explain low-, medium-, and high-technology robots and how they differ.
8. What are the two broad categories of robot end-effectors? Give examples of each.
9. List and explain the various types of contact sensors.
10. List and explain the various types of non-contact sensors.
11. List and explain the four broad categories of robot programming languages.

These case studies were provided by the Society of Manufacturing Engineers (SME). They are excerpted from the SME's *Manufacturing Insights*® series of videotapes. These case studies give students actual examples of the way the advanced manufacturing technologies covered in this chapter are being applied in "real-world" manufacturing settings.

As you read each case study, relate the examples cited to the material presented in the text of the chapter. This combination of textbook information and real-world examples will be particularly valuable to your understanding of advanced manufacturing technologies.

COMDIAL

A pair of inexpensive tabletop robots take on assembly tasks at Comdial and boost production of telephone parts nearly 400%. A one-year payback proves robots can be justified for short-run assembly projects.

Much of Comdial's manufacturing is done in Charlottesville, Virginia. Comdial makes telephones for business and residential use. Products include traditional telephones for the office and home and a variety of unusual telephones for phone users with special tastes.

An important component of many Comdial telephones is the key strip. Production requirements range between 1000 and 2000 per day. The assembly is made up of two primary subassemblies: a plastic key body and a printed circuit board. The two components are joined mechanically. First, the circuit board is fitted over the various pins that protrude from the key body. Then, three of the pins are turned 45 deg each to secure the board. After assembly, a solder mask is applied to specific holes and traces on the circuit board.

In the past, assembly and solder mask application were performed manually. Engineers at Comdial searched for ways to automate the job, but justification was not easy. The potential for savings was limited, so capital investment had to be held to a minimum. Worse yet, the life of the product was expected to be no more than one to two years at most. Expensive special equipment would not pay for itself.

Today, Comdial assembles and applies solder mask in a single step using two Alpha robots supplied by Microbot, Inc. The complete semiautomated system cost only $25,000 and paid for itself in only one year. Furthermore, work that used to take three operators is now done by only one.

The tabletop system consists of a turntable, two Alpha robots, and an operator station. The operator manually loads a circuit board and key assembly into a fixture mounted on the turntable. At the first station, a robot applies consistent force to hold the circuit board against the key assembly, then twists three pins to secure the two components. Solder mask is applied from a dispensing needle guided by a robot at station two. Completed assemblies are unloaded manually.

The Microbot Alpha robot is a relatively new breed of tabletop robot designed for light-duty jobs. Capable of handling workpieces up to 3 pounds, the Alpha robot is driven by precision cables—a less expensive approach than all electric robots. The robot has five axes and can operate at speeds up to 50 in. per second. Repeatability is quoted at 1/15000 in.

When the robot assembly system was first installed, output was about 200 parts per hour—almost twice the production possible using manual methods. To speed up the operation, engineers at Comdial used off-line computer analysis of the robot programs. By reducing dwell time and streamlining all robot motions, output was doubled to nearly 400 parts per hour. Savings are significant. The cost of two operators is avoided. With robotic application, 50% less solder mask is used, reducing materials cost by several thousand dollars per year. Because assembly and masking operations are now performed in one step, in-process inventory is eliminated.

At Comdial, low-cost table-top robots provide the simple solution—the solution that proves that even short-run production can be automated reliably with a good return on investment.

FORD MOTOR COMPANY

The Climate Control Division of Ford Motor Company operates one of the largest and most productive flexible assembly systems in the United States. Robots, hard automation, and manual stations are combined to boost productivity 45% in the assembly of heater and air conditioner controls.

Heater and air conditioner controls used in Ford automobiles consist of a variety of parts, including brackets, washers, pushnuts, lev-

(continued)

ers, screws, panels, and knobs. For years, these parts were joined manually because no economical means of automation could be found. Today, in Plymouth, Michigan, the Climate Control Division assembles two families of heater controls—six different models—using flexible assembly techniques. The new hybrid system assembles over 300 control units every hour for use in automobiles.

Built by Evana Tool and Engineering, Inc., Ford's flexible line combines the advantages of manual assembly, robotic assembly, and automatic testing—all in one special system. The heart of the line is a nonsynchronous pallet conveyor that shuttles fixtures to and from various stations of the system. Operators at six stations load and unload workpieces and perform assembly tasks. IBM 7535 Manufacturing System robots are used at seven assembly stations. Five stations employ dedicated hard automation.

The concept of the system is simple—assemblies are built up progressively as they move through the line, stopping at selected assembly and test stations. Different models stop at different stations, depending on the operations required. About half the stations are common to all models. Manual stations in the system are set up for load and unload chores and assembly jobs that involve components that are susceptible to damage. Manual operations are also used for work that cannot be automated economically or reliably.

The first operation is manual—the worker places a bracket and a bulb and wire subassembly in a fixture, then places a lever on the bracket. In another manual station, the operator flips the control unit upside down, adds a panel and lens assembly, and positions a switch.

Other stations are also tended by operators, including the unload station at the end of the line. If an assembly is rejected for any reason—a missing component or failure at any test station, for example—a label is automatically printed and placed with the unit for rework at a later time.

Robots are used for operations that require flexibility. At one station, a robot places insulating washers on pivot posts. The robot places one or two washers on posts in any of four positions, depending on the model being run. A sensor in the vacuum gripper determines whether the washers are picked up successfully.

At another robotic station, pushnuts are placed on pivot posts. A force sensor monitors resistance. Screws are added at several other robotic stations. At one station, the robot drives two or three screws, depending on the model. The robot is equipped with selective compliance in the horizontal plane to compensate for minor variations in hole positioning. The tapered screws used in the process are presorted and qualified to ensure dimensional accuracy. A robot at another station places knobs on levers. Three, four, or five knobs are assembled, depending on the model.

Hard automation is used for operations that do not vary from model to model. Three different test stations are equipped with special equipment. One test station checks how easily the levers move. Another checks whether the knobs are assembled correctly. An electrical check station ensures that the light bulb is working, then hot date stamps the assembly.

The system is tooled to run any of six control units in random order. However, parts are usually run in batches with changeovers limited to one per shift. Five or six changeovers per shift are sometimes necessary.

Controlled by a Modicon 584 programmable controller (PC), the system is changed over at the push of a button—no changing of fixture details, no adjustments of any kind. The PC handles it all. It stops the fixture carriers at the right stations and tells the robots which programs to run. It even flashes messages at manual stations to tell the workers which models are next. The PC also keeps track of rejects and monitors system performance.

With manual assembly, only four out of five heater and air conditioner controls passed inspection the first time through. With flexible assembly, 97 to 98% pass on the first run. This first attempt at flexible robotic assembly has been so successful that Ford engineers are having a similar system built for a new family of heater controls.

Ford says quality is job one—blending human effort, robotics, and special equipment helps get the job done right.

DELTA FAUCET

Long known as a leader in the design and manufacture of top-quality faucets, Delta Faucet unveiled a unique automated system for the assembly and packaging of faucet components. A special assembly

(continued)

machine and robot-driven packaging station work in tandem to automatically assemble and package more than 50 parts per minute—over 10 million assemblies per year.

Delta Faucet is one of six U.S. and three overseas companies that make up the Plumbing Products Group of the billion-dollar Masco Corporation. Delta Faucet's second largest manufacturing plant is located in Chic-a-shey, Oklahoma. The facility houses a variety of operations, including machining, plating, and assembly and packaging. Average production ranges between 10 and 12 million faucets per year.

A critical component of every faucet is the aerator—the nozzle that introduces air into the stream and straightens the flow of water as it leaves the faucet. A replaceable component, the aerator is threaded to the end of the faucet. Because Delta Faucet produces so many different kinds of faucets, the company uses several different aerators of similar design, varying in size and finish. Traditionally, Delta Faucet purchased millions of aerators from different suppliers to satisfy production schedules.

As part of efforts to reduce costs by bringing the work in-house, engineers at the company developed a new patented aerator. Unlike aerators used in the past, the new design is self-cleaning so clogging is not a problem. The new aerators were also designed with automated assembly specifically in mind. The unit consists of six plastic and rubber components pressed into a cylindrical housing.

Today, instead of buying aerators from suppliers, Delta Faucet assembles and packages the units in-house using a special assembly machine built by Kingsbury Machine Tool Company. Complete with a robotic packaging station, the system allows Delta Faucet to assemble and package aerators in one step without human intervention.

Controlled by an Allen-Bradley 2/20 programmable logic controller (PLC), the indexing table assembly machine is equipped with 16 assembly fixtures, and 7 vibratory bowls for feeding components. Pick-and-place units actually perform the assembly operations.

The automatic assembly process begins with the placement of an aerator housing in a fixture. The fixture is indexed through the remaining assembly stations, where other components are added. The various plastic and rubber components that make up the functioning element of the aerator snap together easily during automated assembly—the parts are held firmly in the housing due to a slight interference fit.

A probe station follows each assembly station. The probing operations verify that each component is present and properly seated. The probing mechanisms are also designed to check dimensional stackups. In some cases, dimensional variations as small as 4/1000 in. can be detected.

At the last station of the machine, parts are unloaded automatically to a belt track leading to a packaging station. A General Electric Model A3 programmable robot is assigned to load the aerators into cardboard boxes to protect the parts in shipping and handling.

Engineers at Kingsbury and Delta Faucet considered using pick-and-place units for the packaging job, but eventually specified a robot for two reasons. The packaging task entails many different motions and logic sequences, so the job would require as many as four pick-and-place units along with related hard automation. With a single programmable robot, floor space requirements are reduced and flexibility is increased.

The operation is completely automatic. First, an empty box is shuttled from a magazine into position for loading. Next, a cardboard separator is placed in the box. A proximity sensor indicates when 10 aerators have accumulated on the belt track. When a full set is present, the robot picks up all ten, moving them into position inside the box. The robot is equipped with a special vacuum gripper. The rows of aerators are nested; every other row of parts is automatically offset by one half-diameter during loading to ensure maximum space utilization inside the box. After eight rows of parts are in place, the robot employs a special rake to make room for the last two rows. When ten rows of parts have been packaged, the robot places a cardboard separator in the box.

This operation is trickier than it seems. The relatively high porosity of the cardboard separators makes it difficult to pick up only one layer using a vacuum gripper. To solve the problem, Kingsbury set up the system so that the vacuum is automatically reduced when the cardboard is handled and a device was installed to strip away unwanted layers as the gripper pulls away from the separator magazine. The packaging operation is completed when ten layers of parts are in place. After a final layer of cardboard is added, the box is ejected to an outgoing conveyor. Capable of assembling and packaging 3120 parts per hour, this usual production system generates savings in many ways. Aerator production at Delta Faucet is less costly than outsourcing.

(continued)

The equipment can assemble a variety of aerators: parts with plated brass housings with external or internal threads or aerators with unplated plastic housings. In fact, the system can assemble the internal plastic and rubber components without any housing. Assembled and packaged just like complete assemblies, these aerators without housings are shipped overseas for final assembly.

Changeover that requires replacement of minor fixture details takes just a few minutes. Program changes for the packaging robot can be accomplished while the system is in operation.

One of the most important benefits is that this unique system can run untended for many hours at night, boosting overall efficiency and assembly for dimensional accuracy at each assembly stage. Overall product quality is improved over that possible using manual assembly methods.

Aerator assembly and packaging in one step is a proven automation advance that keeps the faucets flowing at Delta Faucet.

TELECOM

Canada's largest producer of telephones and related equipment unveiled a robotic assembly cell for terminal connectors that operates at only one-fifth the cost of manual assembly.

Northern Telecom produces a full range of telephones, systems, and switching equipment in plants across Canada and the United States. Annual sales are more than $3½ billion. At the company's Montreal plant, robots are used to trim the cost of assembling terminal blocks. Two different types are required, curved and flat. Both types of terminal blocks are made up in the same way. Ten binding posts are heated to about 500°F, then pressed into the plastic block. Terminal blocks are grouped together inside a terminal housing. Curved blocks are used for aerial housings, the cylindrical terminal units that join sections of overhead telephone wires. The flat terminal blocks are used for splicing underground cable.

In the past, both types of terminal blocks were assembled manually. The operator would load binding posts into cavities of a heated die, then load a plastic block into the upper section. When the binding posts reached insertion temperature, the operator would cycle the press, driving the plastic block over the posts. The thermal sealing process took about 40 seconds. The operator tended two presses, manually loading components and unloading finished workpieces.

Not only was the manual process tiring, it was also hot. Ten years ago, Northern Telecom developed a special machine to assemble the terminal blocks, but it was slow and required too much attention. It did not work out and the company returned to manual operation.

Today, both curved and flat terminal blocks are assembled automatically using a unique robotic cell. It is not faster, but quality is improved and labor costs are reduced. The cell is driven by two Unimate Puma series 560 robots supplied by Unimation Inc. The high-precision, all-electric robots feature six axes and employ Unimation's VAL programming language. Each robots tends three assembly presses, performing identical operations. Five presses are set up to run curved terminal blocks; one is tooled to run flat blocks. Parts are fed to the robots by several vibratory bowl feeders.

The assembly process is simple. First, the robot picks up a pneumatic ejector tool from its holder and proceeds to load ten binding posts into position in the heated die. The binding post ejector tool is similar to a pneumatic screwdriver feeder, but without the screwdriver. An optical sensor in the ejector tool nozzle verifies proper positioning of every binding post. If a post is not seated correctly on the first try, the robot automatically begins moving the tool in a special circular motion until the post drops into position. When all ten posts are seated in the die, the robot returns the ejector tool to storage position. Once the upper die is clear, the robot picks up a plastic block from a vibratory feeder and snaps it into position in the die. The press is cycled automatically.

A variety of sensors are used to verify part presence and position within the cell, including simple microswitches, pressure detectors, and infrared light-emitting diode (LED) sensors. The cell is set up to run untended for as long as four hours before the various vibratory bowls must be refilled. The robots keep track of the failure rate of each press, and if a problem develops, a signal is given to the setup person.

The assembly cell is managed by the robot controllers, a pair of LSI-11 minicomputers by Digital Equipment Corp. Linking the computers allows each to function smoothly without mishap. For example, both robots pick up parts from a central vibratory feeder; if one robot is removing a part from the feeder, the other is programmed to wait.

(continued)

Producing over 800,000 terminal blocks per year, the dual robot assembly cell is credited with savings of more than $50,000 annually. It was designed, built, and programmed entirely inhouse with a straightforward approach, a two-year payback, and 90% uptime, costs are cut at Northern Telecom.

IBM

It's a maze of conveyors and elevators, assembly robots, palletizing robots, testing stations, and computers. It's IBM's fully integrated assembly and packaging line in the Research Triangle near Raleigh, North Carolina. The computer-controlled assembly and packaging facility is about as close to fully automatic as current technology can deliver. Prepackaged components are automatically unpacked and assembled, then repackaged with minimal human intervention.

A new line of IBM computer terminals is modular, consisting of separate monitors, keyboards, and logic units. In the redesign process, the hardware was greatly simplified, keeping ease of assembly in mind. The new logic units consist of only seven parts: two plastic covers, a power supply, a logic card, a key lock, and two screws. The new automated facility consists of two lines that are nearly identical. In total, nearly 20 robots are programmed to assemble and package a family of three logic units.

Primary components of the assembly are supplied by vendors in standard boxes that are palletized and banded according to IBM specifications. Incoming pallets of boxed parts are placed on a pallet conveyor that feeds the automated system. The pallets are automatically debanded, then shuttled to one of four storage conveyors, depending on part type. The stacks of incoming boxes are automatically depalletized by a T3-776 Cincinnati Milacron robot. Individual boxes are shuttled to one of three conveyors, depending on part type. All boxes have easy-to-remove, lift-off lids—like big shoe boxes—so robots can easily gain access from the top. The conveyors lead to robotic assembly stations positioned around an assembly line. Robots at the stations unload boxes and perform assembly operations.

The central element of the system is a CarTrak automatic conveyor supplied by SI Handling Systems. The conveyor transfers dedicated fixtures to and from the assembly and test stations. Top and bottom covers are packed in the same box. The bottom cover is packaged on top because it is the first to be removed. An IBM 7540 robot at station one removes the box top and the bottom cover. Several operations are performed, including label application and laser marking. The boxes at the first station still contain the top covers; they are moved automatically to a storage conveyor for use in the last operation. Another IBM 7540 robot transfers the bottom cover through a checking fixture, then on to the assembly line. The bottom cover travels around the system, stopping at several robotic assembly stations that add the power supply and the logic card. The keylock is assembled manually. The bottom cover, complete with assembled components is automatically tested then transferred to the final robotic assembly station.

The robot at the last station assembles the top cover. Top covers are taken from boxes transferred from the first station. The top and bottom covers snap together easily due to specially designed tabs and slots. A screw is inserted and other operations are performed before the completed assembly is placed in a box and the lid is added. Finished logic units are packaged in the same boxes supplied by the vendors of the top and bottom covers. Boxed assemblies are banded, then palletized and wrapped in plastic. The pallets used here are the same pallets that were unloaded at the beginning of the line. An automated guided vehicle takes the pallets to the shipping dock for delivery to a distribution center.

Many of the robots are controlled by IBM personal computers. Overall system control and management reporting is handled by an IBM Series One computer.

This is the systems approach to automated assembly and packaging—IBM takes the broad perspective—from vendor to distribution and everything in between.

Chapter Six

Major Topics Covered

- Overview of Laser Technology
- Classes of Lasers
- Types of Lasers
- Components of Lasers
- Excitation Mechanisms
- Feedback Mechanisms
- General Uses of Lasers
- Manufacturing Applications of Lasers
- Laser Beam Machining
- Safety Considerations—BRH Requirements
- *Case Study: Allen-Bradley*
- *Case Study: Diversified Manufacturing*

Automated Assembly

Automated Guided Vehicles

CAD/CAM

Numerical Control

Industrial Robots

Lasers in Manufacturing

Programmable Logic Controllers

Flexible Manufacturing

Computer Integrated Manufacturing

Other Related Technologies

OVERVIEW OF LASER TECHNOLOGY

Albert Einstein proved the theoretical possibility of the laser in 1917, but it was not until 1960 that a laser was actually developed. Theodore Maiman created the first working laser when he placed a photographic flash lamp next to a solid cylinder of pink ruby. The term **Laser** stands for **light amplification by stimulated emission of radiation**. Although lasers have been available for some time, at least conceptually, it was not until recent decades that applications were found for them. In fact, in the early development of the technology, lasers were described as a solution looking for a problem. This is no longer the case.

A laser is a device capable of generating light that has unique characteristics that are drastically different from those of conventional light. The laser was originally used in military applications, including range finding, distance measuring, target designation or aiming, and target simulation. Medical science has become a major user of the laser. Lasers are now widely used in cancer treatment, extremely delicate surgical operations, and for a variety of diagnostic procedures. One of the fastest growing applications of the laser is manufacturing.

To understand the role lasers play in modern manufacturing settings, as well as the roles they will play in the future, it is first necessary to understand some fundamentals of laser technology. Since the light produced by a laser differs drastically from conventional light, an understanding of the characteristics of laser light is a good place to begin. The important concepts to understand are light waves, velocity, frequency, wavelength, diffraction, interference, and the operating principles and characteristics of lasers.

Light Waves

Over the years, there have been a variety of theories concerning light. The two of the most prominent are the particle and the light wave theories. The **particle theory** holds that light sources produce tiny particles that bounce off reflective material, thereby producing reflected beams of light. Although this theory was once widely supported in the scientific community, it left too many questions unanswered and was replaced by the **light wave theory**. This theory holds that light travels out from the source in much the same way that ripples or waves emanate from the point where a rock is dropped into still water.

This theory answered most of the questions left unanswered by the particle theory, but not all of them. Consequently, a third theory emerged that encompassed both the light wave and particle theories and holds that light actually has a dual nature. The **dual nature theory** of light holds that light emanates from its source in tiny packets of wave energy that have some of the characteristics of particles of mass. These tiny packets are known as **photons**. This theory says that light is wave energy that travels in a manner similar to particles of mass.

Figure 6–1 is an illustration of a wavelength. The key concepts are amplitude, frequency, period, and wavelength. The **amplitude** is the distance from the centerline to a negative or positive peak on the light wave. **Frequency** refers to the number of electromagnetic oscillations per second of the light wave. Light at any place within the overall spectrum oscillates at a specific fixed rate. The frequency of the light wave determines where in the overall spectrum the light falls. This, in turn, determines the color of the light.

The **period** is the amount of time required for one complete oscillation. The **wavelength** is the distance that light can travel (186,000 miles per second) during the period of the light wave. These concepts are all interrelated and can be shown mathematically:

$$\text{Velocity} = \text{Frequency} \sim \text{Wavelength of Light}$$

It can be seen from this equation that the higher is the frequency of the light wave, the shorter is the wavelength.

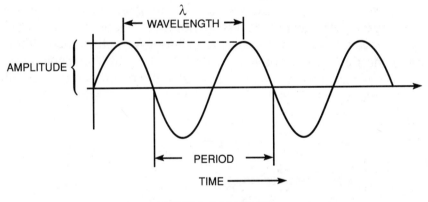

FIGURE 6–1 Light wave.

Diffraction

Diffraction is the way light waves bend as they pass through a narrow opening. This concept is illustrated in Figure 6–2. Note that straight light waves become curved after passing through the small opening, and the opening actually behaves like a light source. Diffraction effects can be easily produced and observed with laser light.

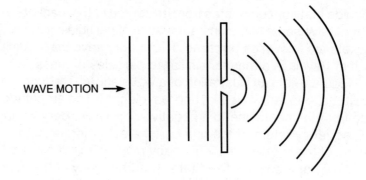

FIGURE 6-2 Diffraction.

Interference

Interference is what happens when two light waves meet at the same place and the same time. There are two types of interference: constructive and destructive. Constructive interference occurs when the crest or trough of one wave meets the crest or trough of another. When this happens, the amplitudes of the light waves come together to form a new light wave with a larger amplitude (Figure 6–3).

When two light waves meet crest to crest, the resultant wave has a larger positive amplitude. When they meet trough to trough, the resultant wave has a larger negative amplitude. When they meet crest to trough, the light waves tend to cancel each other out. The resultant light wave in this case has an amplitude equal to the difference between the two original waves. This is known as destructive interference (Figure 6–4).

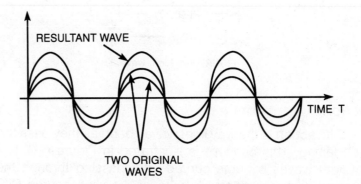

FIGURE 6-3 Constructive interference.

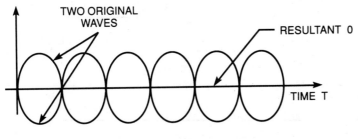

FIGURE 6-4 Destructive interference.

Operating Principles [1]

Before discussing types of laser systems and their applications to machining, a brief explanation of the fundamental principles involved is necessary. As in many advanced processes, the fundamental principle of a laser must be explained at the atomic level.

An atom's orbital electrons can jump to higher energy levels (orbits farther away from the nucleus) by absorbing quanta of stimulating energy. When this occurs, the atom is said to be in the excited state and may then spontaneously emit, or radiate, the absorbed energy. Simultaneously, the electron drops back to its original orbit (ground state) or to an intermediate level. If another quantum of energy is absorbed by the electron while the atom is in the excited state, two quanta of energy are radiated, and the electron drops to its original level. The stimulated or radiated energy has precisely the same wavelength as that of the stimulating energy. As a result, the stimulating energy (pumping radiation) is amplified. This principle is the basis of laser operation. When there are more electrons in the upper energy level than in the lower, the condition is known as population inversion. This condition is necessary for laser operation.

For the laser to convert stimulating energy in the form of coherent light (where all the light waves are in step and intensified to a high power density), certain basic components are necessary, as shown in Figure 6-5 for gas lasers. In Figure 6-6, the light radiated by the excited electrons is in phase with the beam that initiated the reaction. As the now-intensified light continues to travel back and forth through the laser material, more and more electrons are stimulated into giving up their energy, all in phase with the constantly building signal. The light is reflected

---

[1]Excerpted from Chap. 4 of *Nontraditional Machining Processes*, 2nd edition, 1987. *Courtesy of Society of Manufacturing Engineers.*

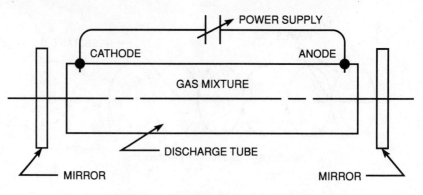

FIGURE 6–5 Typical gas laser configuration.

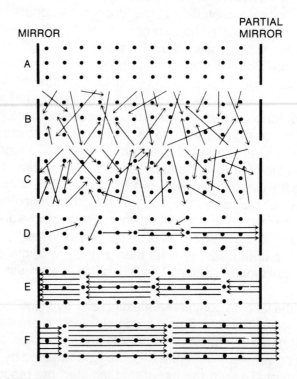

FIGURE 6–6 Nonlasing to lasing. *Courtesy of the Society of Manufacturing Engineers.*

and surges back and forth in the tube. Some of the light is emitted continuously through the partially reflecting mirror. It can then be focused by a lens.

The concepts presented so far are generic. They apply to light in general, including conventional and laser light. Laser light has all of the characteristics of white light, but it also has three additional characteristics that apply only to laser light: (a) **monochromacity**, (b) **divergence**, and (c) **coherence** (Figure 6–7).

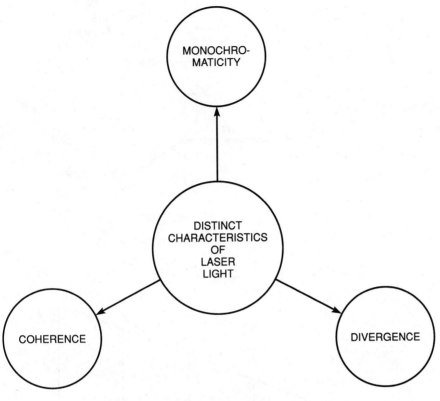

FIGURE 6–7 Distinct characteristics of light.

Monochromaticity

White light contains all of the colors in the spectrum: red, orange, yellow, green, blue, and violet. These colors can be seen when white light is passed through a prism. Laser light, on the other hand, is monochromatic; it has only one color (Figure 6–8). When laser light passes through a prism, it will change direction but it does not break down, indicating that it has only one color.

Divergence

Light tends to spread out as it travels away from its source. This spreading is known as divergence (Figure 6–9). Note in this figure that

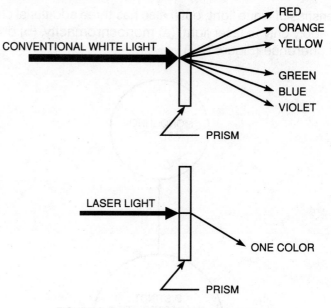

FIGURE 6-8 Laser light is monochromatic.

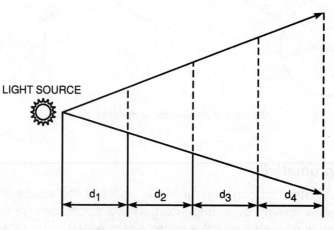

FIGURE 6-9 Divergence of white light.

the farther light travels from the source, the greater is the divergence. Figure 6-9 illustrates divergence characteristic of white light. Laser light has very little divergence: It hardly spreads out at all. A laser beam directed at a wall 1000 ft from the source would form a circle on the wall of less than 12 in. in diameter. A laser beam directed at the moon's surface would form a circle of less than 1 mile in diameter.

Coherence

Laser light is sometimes referred to as coherent light. In coherent light, the light waves are in phase. This means that the crests and troughs match and the wavelength and frequency are equal (Figure 6–10). Note in this figure that all of the key factors—frequency, wavelength, trough, and crest—match. Figure 6–11 illustrates incoherent light.

CLASSES OF LASERS

The laser is one of the most rapidly emerging technologies, but it can be dangerous. Because of this, lasers are placed in four categories according to their power and potential danger: Class I, II, III, and IV.

Class I lasers are the least powerful and the least dangerous. Lasers in this class are not hazardous. **Class II lasers** can be hazardous, but only under certain conditions. Exposure of the eye to Class II lasers could theoretically result in damage. However, damage to eyes is not likely to occur in reality because of the eye's natural blinking reaction,

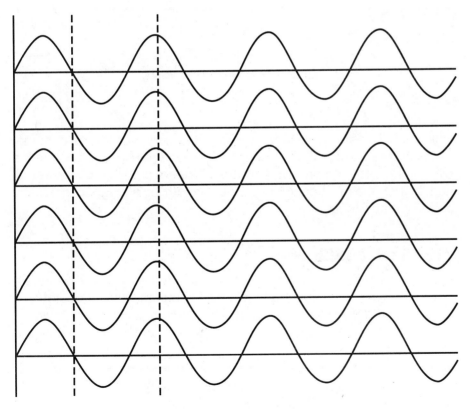

FIGURE 6–10 Coherent light waves.

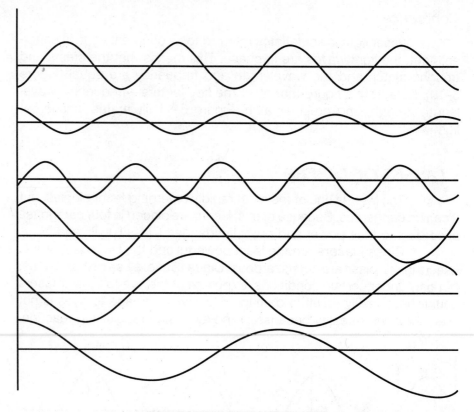

FIGURE 6–11 Incoherent light waves.

which would disturb the laser beam sufficiently to prevent damage. **Class III lasers** are powerful enough to damage the eye even before the blinking reaction can prevent it. **Class IV lasers** are the most powerful, and, in turn, the most dangerous. They are strong enough to ignite any material that will burn.

TYPES OF LASERS[2]

Many media have lasing capability, but only a few lasers are powerful and reliable enough to be practical for machining operations. Lasers can be classified by their lasing medium: solid state, liquid, or gas. The two types used most in machining are the discharge-pumped

---

[2]Excerpted from Chap. 4 of *Nontraditional Machining Processes*, 2nd Edition, 1987. *Courtesy of Society of Manufacturing Engineers.*

carbon dioxide (CO_2) gas laser and the optically pumped solid-state laser.

Gas lasers usually consist of an optically transparent tube filled with either a single gas or a gas mixture as the lasing medium. A typical commercial CO_2 gas laser contains CO_2, helium, and nitrogen (N_2). Carbon dioxide supplies required energy levels for laser operation and nitrogen provides intracavity cooling. The pumping source is some form of electrical discharge applied by electrodes.

The CO_2 laser operates at a wavelength of 10.6 microns either pulsed or in a continuous wave. In high-power gas lasers, the available output is limited by two factors: ability to cool the gas and ability to properly stabilize the gas discharge. Thermal energy upsets the lasing equilibrium of the gas. Cooling is necessary primarily to keep the lower laser level depopulated.

Solid-state lasers consist of a crystalline or glass host material and a doping additive to provide the reservoir of active ions needed for the lasing actions. The original solid-state lasers used ruby (with approximately 0.05% chromium dopant) as the lasing medium.

Another common solid-state laser uses a single crystal of yttrium aluminum garnet (yag) doped with neodymium as the lasing medium. The neodymium-doped yag (Nd-yag) laser is relatively efficient, allows for high pulse rates, and can be operated with a simple cooling system. Both ends of the rod (lasing medium) are parallel and polished to high flatness. The Nd:yag solid-state laser operates at a wavelength of 1.06 microns. Most solid-state lasers operate only in the pulsed mode; however, the Nd:yag laser may be operated either pulsed or in a continuous-wave mode.

Since the Nd:yag laser material is electrically insulating, it must be powered by a means other than a simple electrical excitation. Energy is injected into the laser medium to generate an intense light flux. The light is absorbed by the medium and collimated into a laser beam. The pump that optically excites the laser materials is usually a krypton- or xenon-filled arc-discharge lamp. Figure 6–12 is a diagram of a solid-state laser system.

To efficiently use the light produced by the lamp, the laser rod and arc lamps are mounted in a reflective cavity. The cavity may be in the form of an elliptical or circular cylinder to focus the light from a linear arc-discharge lamp onto the laser rod.

The energy source for the pulsed flash lamps usually consist of a dc power supply and a bank of energy storage capacitors. The capacitors are charged to a predetermined voltage. Their stored energy is then discharged through an inductor to limit current peaks, then through the flash lamp to create the necessary pump light.

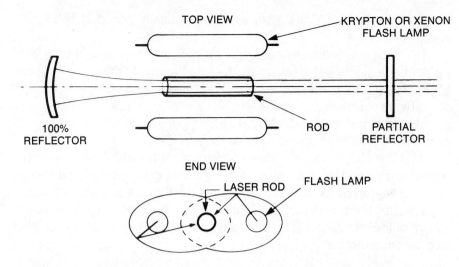

FIGURE 6–12 Solid-state laser systems. *Courtesy of the Society of Manufacturing Engineers.*

In solid-state lasers, removal of waste heat (the excess energy not usefully converted into laser radiation) is a fundamental problem. The radius of the rod is limited by the need to conduct surplus heat to its cooled periphery. This requirement sets a practical upper limit to the power that can be extracted per unit length of rod and thus per system. Waste heat may be taken up by a heat sink in relatively small systems. Larger systems may use streams of fluids to carry waste heat away from the laser cavity. Typical fluid cooling systems are water, water to air, water to water, and refrigerated recirculating water.

COMPONENTS OF LASERS

Four components must be present to have a laser: **active medium**, **excitation mechanism**, **feedback mechanism**, and **output coupler** (Figure 6–13).

Active Medium

To understand this component, it is necessary to first understand the concept of stimulated emission. Stimulated emission is a reaction that takes place when a photon of the proper wavelength comes near an excited atom. This creates another photon. This new photon and the original photon have the same wavelength, are coherent, and begin to travel in the same direction.

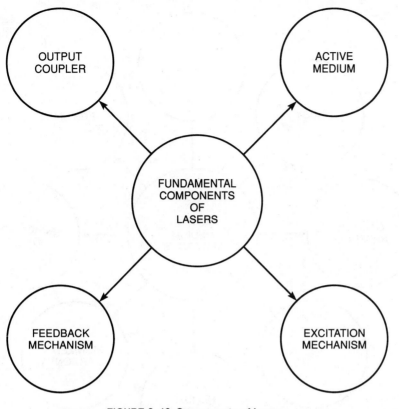

FIGURE 6–13 Components of lasers.

The active medium is the medium in which stimulated emission takes place. The active medium can be solid, liquid, or gas. It can even be a semiconductor material. For example, the active medium for the first laser created in 1960 by Theodore Maiman was a ruby (solid). Some of the more commonly used materials for the active medium component are the following (Figure 6–14):

- Ruby
- Glass
- Erbium
- Organic Dyes
- Ionized Argon
- Helium-Neon
- CO_2
- Gallium Arsenide

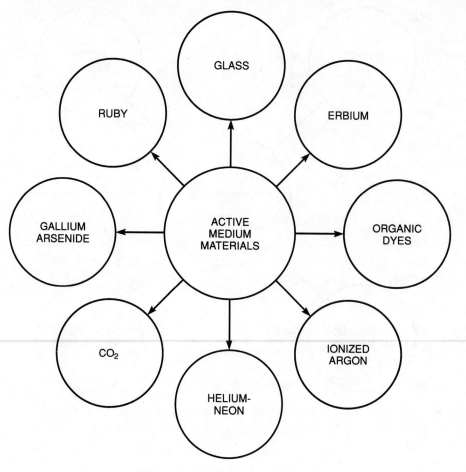

FIGURE 6-14 Active medium materials.

EXCITATION MECHANISMS

Recall that for stimulated emission to take place a photon must approach an excited atom. The excitation mechanism is a device used to induce an excited state in an atom. Three different types of devices can be used as excitation mechanisms.

1. optical
2. electrical
3. chemical

The excitation process involves pumping energy into the active medium. This process is sometimes called **laser pumping**. Each of the three types of devices listed above can be used to pump a laser.

Optical Devices

Optical devices excite the active medium (pump the laser) using light energy. Light energy of a predetermined wavelength from the sun, a flash lamp, or a variety of other sources is introduced into the active medium. Lasers with solid-based active media (i.e., ruby, erbium, glass) use optical devices as their excitation mechanisms. Optical devices are not normally used with active media that conduct electricity (i.e., gallium arsenide, indium arsenide).

Electrical Devices

Electrical devices are used as the excitation mechanism in lasers with an active medium that conducts electricity. Such active base materials as gallium arsenide, gallium antimonide, Indian arsenide, CO_2, ionized krypton, helium-neon, and ionized argon call for electrical excitation mechanisms.

Gas-based active media carry atoms, ions, and electrons collectively known as plasma. When a high-voltage charge is introduced into such an active medium, energy from the current is transferred to the plasma, resulting in excitation.

Chemical Devices

This category is the least commonly used type of excitation mechanism. The primary application of lasers that use a chemical excitation device is in military uses. Energy can be produced by mixing selected chemicals in the appropriate proportions. Energy is released from the chemical bonds that are either made or broken. When these reactions occur within an active medium, the energy produced is sufficient to induce excitation. Large military-oriented lasers such as hydrogen-flouride lasers use chemical excitation devices.

FEEDBACK MECHANISMS

The final fundamental component of the laser is the feedback mechanism. Mirrors are used for this purpose. Feedback is the process of reflecting light produced in the active medium back into the medium. By placing the mirrors properly, light can be reflected back and forth through the active medium. This continual reflecting of light waves back and forth through the active medium keeps stimulated emission at a maximum. The more distance the light waves travel within the active medium, the longer they stay with the medium, and, as a result, the stronger is the amplification of the light. Figure 6–15 illustrates how curved mirrors

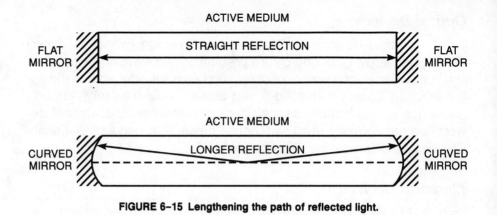

FIGURE 6-15 Lengthening the path of reflected light.

can be used to lengthen the path of reflected light within the active medium.

GENERAL USES OF LASERS

Lasers are constantly finding new applications. One of the areas in which lasers are becoming widely used is manufacturing. However, there are many nonmanufacturing applications of lasers with which students of manufacturing processes should be familiar. The four most common nonmanufacturing applications of lasers are medical, military, communications, and consumer products (Figure 6-16).

Medical Uses

Medical science was one of the first fields to recognize the potential of lasers and to take full advantage of that potential. The first clinical use of the laser was in eye surgery. Because of the laser's special qualities, it proved to be a valuable tool in repairing detached retinas, a very intricate surgical procedure. The laser is now used in a number of eye surgery procedures including photocoagulation, glaucoma relief, and cataract removal. Other medical applications of the laser are regular surgery, endoscopy, photoradiation, and research.

Military Applications

Military applications of the laser include range finding, target designations, training simulation, and target destruction. By directing a laser beam on a target and measuring the amount of time it takes to be reflected back, military personnel can determine the range of a target.

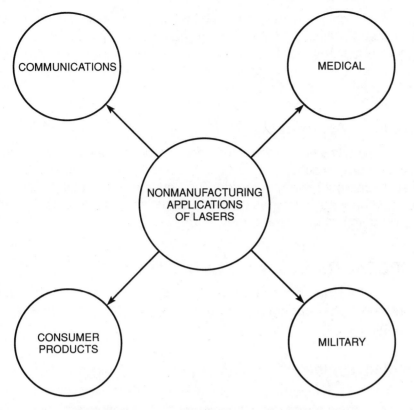

FIGURE 6–16 Nonmanufacturing applications of lasers.

Target designation involves using a laser to illuminate an object, thereby designating it as a target and guiding a missile or other projectile to it.

Training simulation involves using the laser as an aid in making military training more realistic. One of the more widely used approaches is to direct a laser beam at a target that is equipped with a detecting device. When the detecting device is triggered, the target has been hit.

The military application most people readily think of is the laser used as a weapon. However, in reality, this is the application that is the least developed. Lasers have been used as surface-to-air weapons on an experimental basis. Ground, air, and space weapons based on the laser are in various stages of development at present.

Communication Applications

Communications technology is a field that represents fertile ground for laser applications. One of the most rapidly emerging communications of the laser is data transmission. In transmitting data electronically, the higher the frequency of the carrier signal, the more data it can carry. Laser light has a very high frequency; its frequency is 100,000,000 times greater than that of the highest FM radio signal. By joining a laser with fiber optic cable, data transmission can be increased significantly.

Consumer Applications

At some time in the future, consumer applications might represent a large application area. At this time, consumer applications of the laser are only beginning to emerge. They include home security systems, reading holographic images on credit cards, and reading optical codes printed on consumer products.

MANUFACTURING APPLICATIONS OF LASERS

The major manufacturing applications of lasers are machine alignment, heat treating, bar code reading, marking, and laser beam machining.

Machine Alignment

The laser enables manufacturers to produce parts that require greater degrees of accuracy than normal. In fact, lasers are able to align machines to such high degrees of accuracy that current calibration standards no longer suffice. Prior to the use of laser alignment, an accuracy of 0.0001 in. was considered very good. With laser alignment, accuracy of 0.00001 in. is possible.

Heat Treating

Heat treatment is one of the more commonly used manufacturing processes. Metals are heat treated to harden them. At this point in the development of lasers, traditional heat treatment processes are more economical. However, lasers can be used to treat small areas that are hard to reach using traditional methods. If a large machined part has a small area on it that needs to be hardened, a laser can do the job better than traditional processes. Gear teeth are another application in which the laser is the more appropriate choice.

Welding

Laser beam welding (LBW) is a fusion process used to join metal pieces as thick as 1 in. The laser beam directed on the metal pieces melts the edges, causing the pieces to fuse together. No filler is necessary with LBW as is sometimes the case with conventional welding processes. However, LBW can use filler if appropriate. Laser beam welding can be used for structural assembly, sealing, and conduction welds. Both seam and spot welds can be accomplished with LBW.

Figure 6–17 illustrates one of the most widely used techniques: keyhole welding. Note from this figure that, heating with keyhole LBW, the laser beam cuts a deep hole into the material to be welded. As the laser lases in one direction, the beam melts the material directly ahead of it. The molten metal works its way around the beam to a location behind it, where solidification occurs.

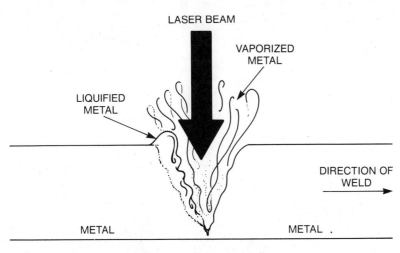

FIGURE 6–17 Keyhole welding.

Marking

Many manufactured parts must be inscribed with a name, number, or some other type of alphanumeric code. Laser marking has proven to be a fast, effective way of inscribing products. There are two widely used approaches to laser marking. In the first method, a special die or mask containing the desired masking pattern is placed between the reflecting mirrors of the laser. The laser beam must pass through the mask. As it does, it takes on the shape of the pattern, and, in turn, inscribes the part with the pattern.

In the second method, a laser beam is deflected by computer-controlled mirrors. One mirror controls the X axis. By controlling these mirrors properly, the laser beam can be deflected in such a way as to inscribe a part. This process is illustrated in Figure 6–18 and an example of a part inscribed using laser marking is shown in Figure 6–19.

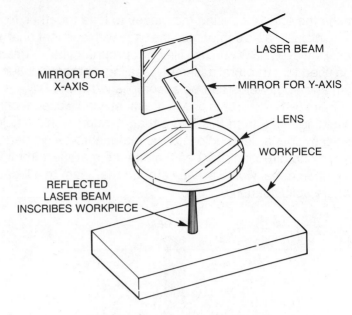

FIGURE 6–18 Laser inscribing system.

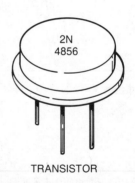

TRANSISTOR

FIGURE 6–19 Laser-inscribed part.

LASER BEAM MACHINING

Laser beam machining is one of the principal applications of the laser in manufacturing. Laser beam machining involves using lasers to cut and drill materials. This section provides a general description of the laser beam machining process and specific descriptions of drilling and cutting.

Process Description[3]

The laser yields a power density sufficient to vaporize any known metal and even diamond. It can readily cut nonmetals including wood, cloth, paper, and advanced composites.

Not all of the material is removed by evaporation, however. Laser machining is basically a high-speed ablation process as illustrated in Figure 6–20. The workpiece is heated so that surface melting occurs. The evaporation of a very small portion of the liquid metal takes place so rapidly under the high intensities of a focused laser beam that a substantial impulse is transmitted to the liquid. Material leaves the surface not only through evaporation, but also in the liquid state at a relatively high velocity. In some cases, cutting may be enhanced by introducing a gas such as oxygen to the point at which the laser beam is working.

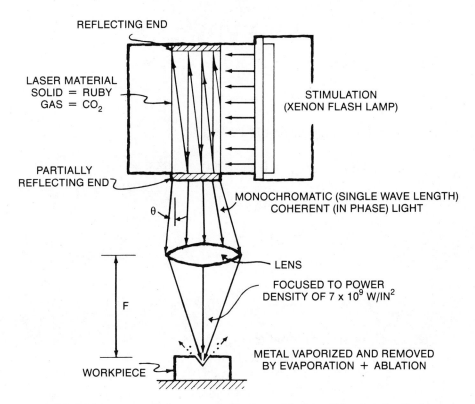

FIGURE 6–20 Laser machining process. *Courtesy of Society of Manufacturing Engineers.*

[3]Excerpted from Chap. 4 of *Nontraditional Machining Processes*, 2nd edition, 1987. *Courtesy of Society of Manufacturing Engineers.*

Machining a hole with a laser beam requires a short, high-intensity pulse. The amount of energy needed to vaporize a volume of material can be calculated. This is approximately the energy required to raise the metal to its vaporization temperature, plus the latent heat of fusion and vaporization as shown in the computation that follows:

For example, to drill a hole 0.45 in. (11.4 mm) in diameter in a 0.04-in. (1.0 mm)-thick sheet metal by means of a laser, the volume of the metal to be evaporated equals

$$6.4 \times 10^{-5} \text{ in.}^3 \cong 1 \text{ mm}^3 \text{ or } 0.008 \text{ g.}$$

The energy required for vaporization of 1.0 g of the metal requires

a. heating it from room temperature to melting point:

$$E_1 = C(T_m - T_o) = 0.11 \ (1535 - 20) = 167 \text{ cal,}$$

b. changing it from solid to liquid at T_m:

$$E_2 = L_f = 65 \text{ cal,}$$

c. heating it from melting point to boiling point:

$$E_3 = C(T_b - T_m) = 0.11 \ (3000 - 1535) = 161 \text{ cal,}$$

d. changing it from liquid to vapor at T_b:

$$E_4 = L_v = 1630 \text{ cal}$$

and

$$E_1 + E_2 + E_3 + E_4 = 2023 \text{ cal} = 8500 \text{ J}$$

where: C = specific heat (cal/g)
   T_o = ambient temperature (°C)
   T_m = melting temperature (°C)
   T_b = boiling temperature (°C)
   L_f = heat of fusion (cal/b)
   L_v = heat of vaporization (cal/g)

Thus, vaporization of 0.008 g (1.0 mm³) of the metal requires approximately 68 J.

Assuming that it requires a laser energy on the order of 100 J and also assuming a pulse length of 10^{-5} sec, the required power would be

$$\frac{10^2 \text{ J}}{10^{-5} \text{ sec}} = 10^7 \text{ W.}$$

To meet the basic requirements for industrial application, laser systems must meet the following specifications: (1) sufficient power output, (2) controlled pulse length, (3) suitable focusing system, (4) adequate repetition rate, (5) reliability of operation, and (6) suitable safety characteristics.

Proper consideration in product design and selection of materials for parts to be processed by laser can yield major cost and quality benefits. Laser processing provides its own unique opportunities to simplify product design and to select materials from which the products are to be made. Laser beam processing may not require as flat a surface, perpendicular to the laser beam, as the surface geometry required for mechanical drilling. Extra operations to provide such flat surfaces, if not required for other purposes, may at times be eliminated.

The laser may be used to drill, cut, mark, heat treat, or in some cases, even weld finished or nearly finished parts. The laser process often may be used quite close to other elements since the heat-affected zone is small. Care must be exercised to protect finished surfaces and other parts, including machine parts and tooling, from splashing slag. Processing some parts and materials may require a surface coating to provide a uniform, highly absorptive surface for the far-infrared laser energy (primarily for CO_2 lasers). Such coatings may be effective up to the melting point of any metal. Coatings are also used to provide high absorptivity at acute angles of laser beam incidence.

Parts that are to be laser heat treated to increase surface hardness may often be designed to be made from low-cost, easy-to-machine, plain carbon steels rather than alloy steels.

In all cases where laser processing is used, part designers must take into account that energy density input is high. Although the heat-affected zone is shallow, very high thermal stress may be applied and high levels of residual stress due to processing also may be present. These stresses are due to the rapid heating and cooling rates and the transformations that occur in the structure of the material. Sharp corners, sharp reentrant angles and rough machined surfaces all compound stress problems. Often there is a need for post-laser processing.

A wide variety of fixturing and positioning systems and options including those for beam splitting, beam direction, and spot location are available. Numerical control equipment is widely used in conjunction with the laser. When several parts must be addressed or the work must be addressed from several directions, it is often necessary to split the beam into two or more parts.

The most economical and reliable alternative to beam splitting is the multibeam laser. Such a laser has a resonator that emits two or more beams. There may be two or four lasers inside the cabinet. Each laser utilizes a common power supply, electronic controls, and gas recycler to make a very efficient system to address more than one work station.

For applications where a pattern is to be engraved, a surface is to be hardened, or a pattern of holes is to be drilled, a variety of systems are available that establish the track the laser beam is to follow. In some cases, a photo-etched or reflective mask may be used to shield some portions of the part while the beam is permitted to impinge on the part where desired.

Laser power output can be altered by various means. The laser may be scanned over a given area to average the energy flux or to control the heat input required for heat treating, scaling, or identification marking.

Parts must be accurately positioned in relation to the axis of the laser beam, and the laser beam must be focused properly. A laser system including the laser focusing and viewing optics for a system is illustrated in Figure 6–21. Optics are used for focusing the working beam. A viewing system may not be needed for a setup dedicated to certain tasks, but is essential for a manual, versatile-use setup. Visual inspection of welds is very important to the operator for making in-process adjustments. The use of closed-circuit television (CCTV) monitor is less eye-fatiguing during production than viewing through a microscope.

Applications[4]

The two principal laser machining processes are drilling and cutting. Laser processing is not usually employed as a mass material removal or heating process. Very small amounts of material are affected by each pulse of the laser. Many closely spaced pulses within a short time accomplish the desired results.

---

[4]Excerpted from Chap. 4 of *Nontraditional Machining Processes*, 2nd edition, 1987. *Courtesy of Society of Manufacturing Engineers.*

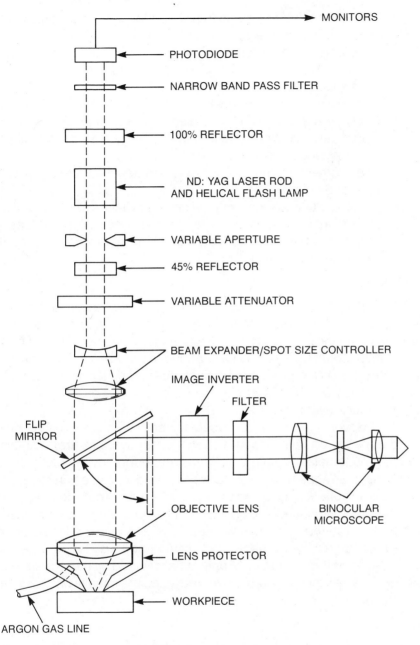

FIGURE 6-21 Laser system.

Application of lasers in material processing, if it is to be successful, requires good coupling of laser energy to the part. Laser processing is essentially a controlled heating process. The most important material properties and characteristics are as follows:

1. Those affecting the manner in which the light is absorbed by the material. These are reflectivity of the surface at the particular wavelengths being used and the absorption coefficient of the bulk material. The ability of a material to absorb the infrared radiation from a CO_2 laser can be obtained from electrical conductivity tables because absorption is inversely proportional to electrical conductivity. Materials with good electrical conductivity such as gold, copper, and aluminum are poor light energy absorbers, while plastics and wood are almost perfect absorbers.
2. Those governing the flow of heat in a material—thermal conductivity and diffusivity.
3. Those relating to the amount of energy required to cause a desired phase change—density, specific heat (actually heat capacity and the latent heat effect, i.e., heat of fusion and heat of vaporization.

Drilling

Drilling holes was one of the early uses of the laser beam. Not very useful, but one use that attracted attention to the process, was the punching of holes in razor blades. Since that time, laser drilling has become more sophisticated. Holes continue to be drilled in hardened steel and are now drilled in all metals (including many high-temperature alloys), plastics, paper, rubber, ceramics, composites, and crystalline substances such as diamond.

Several methods are used for cutting holes. The technique used depends on hole size and shape. Laser systems are available for cutting round holes, welding, and making perforations in a circular pattern. These devices typically rotate a focusing lens in a horizontal plane (or the plane of the lens) on an axis coincident with the incoming stationary beam. The focused spot will always be on the focal axis of the lens and will be rotated in a circle with the lens. The rotating lens assembly is typically driven by a variable-speed motor and may be equipped with a gas jet. The effective radius of operation is limited by the lens size used. For holes larger than 1 in. (25.4 mm) diameter, it is usually better to rotate the mirror.

Cutting

Lasers will cut hard or soft, tough or brittle, and stiff or resilient materials. It does not matter whether the structure is strong or weak. Laser cutting is usually assisted by a flow of gas, which performs several functions such as cooling the area around the cut and blowing away swarf and slag. Reactive gases increase cutting speed. Self-burning of the material, however, can cause poor kerf quality.

The oxygen supplied in gas jet-assisted cutting enhances the cutting of steel for several reasons. Oxygen reduces reflectivity, helps initiate the exothermic reaction once the metal reaches a high temperature, cools the material around the working area, directs molten metal and vapor away, and sweeps away molten slag from the bottom of the cut.

It is important that the proper gap between the gas jet and the workpiece be maintained. This may be accomplished with self-adjusting, height-sensing units that control the gap automatically regardless of surface unevenness.

Motion control of the laser beam in relation to the workpiece is usually accomplished by an optical tracer or a numerical control system. Numerical control systems vary widely in their sophistication. Some simply control X-Y axis motions, while others may be coupled with a computer system that assists in part layout on a sheet in order to maximize material utilization, establish the cutting path based on dimensional and geometric input, determine stock status or availability of most economical sheet stock, establish shop loading schedules and report job status, maintain cost data, and prepare shipping and billing papers. Figure 6–22 is an example of a three-axis computer numerical control (CNC) laser system.

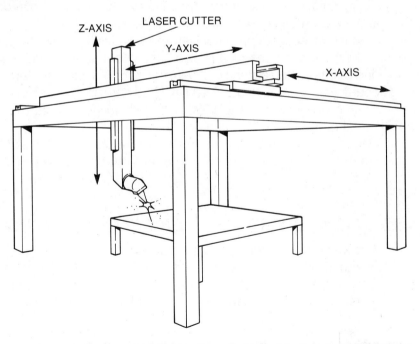

FIGURE 6–22 Three-axis CNC laser system. *Courtesy of the Society of Manufacturing Engineers.*

The laser is capable of producing a very narrow kerf. Operating a laser in the TEM_{00} mode provides narrow kerf widths, square kerf sides, a narrow heat-affected zone, and minimal slag. Kerf widths of 0.005 to 0.015 in. (0.13 to 0.38 mm) have been consistently achieved with the focal point positioned about one-third of the material thickness below the surface using a 0.050-in. (1.25-mm)-diam gas jet nozzle, positioned about 0.015 to 0.020 in. (0.38 to 0.5 mm) above the surface.

All metals are relatively good (80%+) reflectors of the 10.6-micron wavelength energy at room temperature. Reflectivity is also affected by the finish on the material. In the case of exceptionally good reflectors such as aluminum and copper, there may not be sufficient laser energy absorbed by the material to initiate vaporization or even melting. It has been predicted by theory and documented by experiment that a substantial increase in the absorptivity of metals occurs at elevated temperatures. The reduced thermal conductivity of metals in the molten state is considered to be an important factor. In theory, molten aluminum, copper, and silver approach the absorptivity of stainless steel at room temperature. No data are available on the absorptivity of the two-phase mixture of molten metal and its vapor; experience and some approximate calculations indicate that absorption may be as high as 50%. This increase of energy absorption over theoretical makes cutting of aluminum possible. In practice, this also means precise focusing. The above discussion is especially relevant to uses of CO_2 lasers.

Focusing becomes progressively more critical for a given material as its thickness is increased. It is also critical for materials with very high melting points. When cutting 0.020-in. (0.51-mm)-thick tungsten, focusing accuracy must be within ±0.003 in. (0.08 mm). In the case of titanium of the same thickness, it is ±0.020 in. (0.51 mm). [These data apply to a 2.5-in. (64-mm) focal length lens.]

Paper and similar sheet materials can be cut or perforated at very high rates with lasers. Another weak material that can be cut is polyurethane foam.

SAFETY CONSIDERATIONS—BRH REQUIREMENTS[5]

Laser safety requirements are regulated by the Bureau of Radiological Health (BRH), which is a division of the Food and Drug Administration. During the past several years, this bureau has prescribed requirements that all laser manufacturers must follow so that lasers

---

[5]Safety considerations are courtesy of Coherent, Inc.

will have safeguards for personnel. These safeguards include the following:

1. Plasma tube shields: These prevent ancillary laser emissions from the laser tubes.
2. Power supply shields: These are required around the high-voltage power supply to prevent the emission of "soft" X rays.
3. Danger labels: A label must be affixed to every laser, listing its power level and clearly warning of the possible danger involved.
4. Emission indicator: A bulb is lighted near the beam exit point when a beam is emitted from the enclosure.
5. Shutter light: A light is provided near the beam exit to show that the shutter is open and a beam is being emitted.
6. Time delay: A minimum 10-sec time delay is provided in every laser so that if the laser is accidentally turned on, a beam will not be emitted immediately.
7. Safety beam enclosure: This prevents accidental exposure to a beam while turning the output optics.
8. Keylocked controls: These prevent use by unauthorized personnel.
9. Exhaust system: Many materials that are processed give off harmful or toxic fumes when burned. An exhaust system removes these fumes from the operator's environment.
10. Safety glasses: The CO_2 laser beam is stopped by any plastic or glass safety lens. Operating personnel should wear glasses as they would when operating a lathe. Other materials are available to stop other types of laser beams.
11. Authorized personnel: The people who operate the laser system should be trained in the system and trained in these safety considerations.
12. Respect: The laser and system should be treated with the respect that is due to any potentially dangerous machine tool.

KEY TERMS

Laser
Particle theory
Light wave theory
Dual nature theory
Photons
Amplitude
Frequency

Period
Wavelength
Diffraction
Interference
Monochromacity
Divergence
Coherence

Class I, II, III, and IV lasers   Machine alignment
Active medium   Heat treating
Excitation mechanism   Bar code reading
Feedback mechanism   Marking
Output coupler   Laser beam machining
Laser pumping

Chapter Six REVIEW

1. What does the term laser stand for?
2. Describe the development of the first laser.
3. Describe the particle and light wave theories of light.
4. Describe the dual nature theory of light and contrast it with the two other major theories that preceded it.
5. Sketch a light wave in such a way as to illustrate the following concepts: frequency, period, and wavelength.
6. Define the following terms:
   Frequency
   Period
   Wavelength
7. Describe and contrast the four classes of lasers.
8. List and explain four essential components of lasers.
9. List and explain four nonmanufacturing applications of lasers.
10. List and explain the most widely used manufacturing applications of lasers.

These case studies were provided by the Society of Manufacturing Engineers (SME). They are excerpted from the SME's *Manufacturing Insights®* series of videotapes. These case studies give students actual examples of the way the advanced manufacturing technologies covered in this chapter are being applied in "real-world" manufacturing settings.

As you read each case study, relate the examples cited to the material presented in the text of the chapter. This combination of textbook information and real-world examples will be particularly valuable to your understanding of advanced manufacturing technologies.

ALLEN-BRADLEY

Laser gauging provides closed-loop control of a precision grinding operation, and laser marking performs flexible product identification in Allen-Bradley's automated assembly operation.

Milwaukee-based Allen-Bradley is well known to manufacturing engineers as a maker of relays, motor controls, contactors, programmable logic controllers, and highly sophisticated factory local area networks. When Allen-Bradley went head-to-head in world markets for contactors and relays, the company designed a new product. The decision was also made to build these new products on a completely new, fully automated line. Competitiveness in world markets was important for the new line. These contactors and relays were a departure from Allen-Bradley's NEMA designs. Designed to the International Electro-Technical Commission (IEC) standards, the products compete in world markets with products manufactured offshore.

Built in a thoroughly engineered "world contactor" factory-within-a-factory on the eighth floor of Allen-Bradley's Milwaukee headquarters, the new components are fabricated and assembled entirely under computer control. There are only two doors in the facility. Plastic resins, sheet metal, a few dozen standard components such as screws and terminal plates, blank shipping labels, boxes, and cartons come in through one door and finished products in cartons go out the other door. All of the equipment on the line was designed by Allen-Bradley, using a variety of automation engineering principles. Well over 60% of the machines were built in-house, with the remainder built to Allen-Bradley specifications.

(continued)

All of the components in the contactors are made with high precision, a prerequisite for totally automatic assembly. One of the most precise parts is the armature.

After collating and riveting, the armatures travel to a precision grinder. A vapor-and-air cleaning follows the grinding process, to remove any particles. Then the parts are conveyed to the laser gauging unit built by Zygo Corporation.

Essentially, the Zygo unit measures shadows cast by the armature. The advantage of using laser light over other light sources is that laser light is coherent. No refraction will occur over the bar or armature.

The cylindrical gauge bar makes single-point contact tangent to the arms of the E-shaped armature. Overall arm height and any end-to-end variation can be interpolated from readings above the gauge bar.

Individual armatures that are out of specification are kicked into a reject buffer. The Zygo unit also tracks trends, and if several off-specification parts in a row show the same problem, then the Zygo directs positioning of the grinding wheels. Thus, if grinding wheel wear is the cause of the problem—as it is in most out-of-specification grinding—the grinding unit operates under closed-loop control. Should the parts remain out of specification, as might happen with a broken grinding wheel, an alarm sounds, and the section shuts down to await operator assistance. An overhead readout translates numerical error codes from the grinding and gauging units into clear messages.

Armature grinding is among the first operations in making the contactors. The last operation involves laser printing.

Printing or hot stamping the contactors and relays would not be economically feasible. Each of the 125 different models has its own unique contact codes and electrical diagrams. The line is capable of lot sizes of one to many thousands; getting the right printing plates in place would be impossible. It is a job for flexible automation.

A Control Laser InstaMark laser engraving system was selected. A computer determines the numbers and letters that go onto each of the plastic cases. Each contactor has a bar code label and at the laser printer, a bar code reader tells the computer which program to use. A single neodymium-YAG laser beam is manipulated to mark the plastic cases.

Lasers gauging and printing play an important role in totally automatic manufacturing at Allen-Bradley.

DIVERSIFIED MANUFACTURING

At Diversified Manufacturing in Lockport, New York, calibration and alignment procedures once took days; now they take only hours, thanks to precision laser alignment and calibration. Laser Line Inc., Diversified's subsidiary, offers laser alignment and calibration services using Hamar laser alignment systems and the Hewlett-Packard laser measurement system. Unlike high-powered laser systems that heat, melt, and vaporize, these systems feature low-powered helium-neon lasers.

Image selected from SME's Lasers In Manufacturing video from the Manufacturing Insights® Video Series.

(continued)

Laser alignment and calibration systems use the inherent stability of the laser beam to provide accurate measurements, up to 0.0001 in., over distances that would otherwise be impractical.

While a full calibration cycle by traditional methods can take up to two weeks, calibrating and aligning a machine tool takes only about 16 hours with laser technology. Laser Line's Hamar alignment system can detect and measure errors in these geometric aligning applications:

- Straightness checking
- Machine leveling
- Flatness checking
- Column squareness and perpendicularity
- Spindle alignment

The Hamar system works on the principle of "dynamic testing." The laser beam is adjusted to be parallel with the motion of the machine tool by bringing the "target" and the laser head into precise alignment. The target contains a two-axis quadrented continuous cell that is used to find the center of the laser beam. When both the beam and target are aligned, the system shows no error. If the center of the beam and the center of the target are offset, either horizontally or vertically, the system displays the error.

Once the target and the laser head are aligned, the target is moved along the line of motion. The readout unit displays any geometric errors. Perpendicular alignment and squareness can be checked by adding a squareness optic and aligning the laser beam with a target attached to the machine's spindle nose. The optic bends the beam upward so it is perpendicular to the laser's horizontal beam. Squareness of the machine's vertical travel in relation to its horizontal travel is displayed as the column moves up and down.

This setup can also be used on bridge-type machines, since the squareness optic also sweeps a 360-deg flat plane. Flatness and squareness are checked by moving a target to different positions on the right and left column ways and the ways of the overhead bridge.

The Hamar System can also be used to check the positioning accuracy and repeatability of rotary tables. This setup uses an optical polygon that is positioned at the center of the plane of rotation with the target positioned next to the laser. Since the readout displays the error in thousandths, the distance between the target and the surface of the polygon is calculated so that one one thousandth equals one arc second.

Alignment of opposing spindles, spindles to master parts, tailstocks to headstocks, and parts to spindles is also simplified with the Hamar spindle laser. Mounted into the spindle, the laser projects the axis of rotation, while the target, mounted into the opposing spindle or part, reads the misalignment.

In addition to the Hamar alignment system, Laser Line uses the Hewlett-Packard laser measurement system for dimensional measurements and calibration. Hewlett-Packard's laser interferometer uses various optics to detect and measure errors in the following:

- Distance and velocity
- Pitch and yaw motions
- Straightness
- Squareness

The two-frequency laser interferometer measures distance by detecting the shift in frequency between a moving optic and a stationary optic. Any point can be defined as the measurement starting point and designated as "zero." For example, the interferometer checks distance and velocity by comparing the frequency of the stationary beam with the frequency of the beam that is reflected from the moving optic. The real-time display shows the amount of error as the machine comes to each command point.

Major Topics Covered

- Programmable Logic Controller Defined
- PLCs Versus PCs
- Historical Development of PLCs
- Configuration of the PLC
- Operation of the PLC
- Applications of PLCs
- *Case Study: Molex Corporation*
- *Case Study: Babcock & Wilcox*

Chapter Seven

Automated Assembly

Automated Guided Vehicles

CAD/CAM

Numerical Control

Industrial Robots

Lasers in Manufacturing

Programmable Logic Controllers

Flexible Manufacturing

Computer Integrated Manufacturing

Other Related Technologies

PROGRAMMABLE LOGIC CONTROLLER DEFINED

Programmable logic controllers (PLCs) are at the very nucleus of automated manufacturing technology. The Allen-Bradley Systems Division, a leading supplier of PLCs, defines the PLC as "a solid-state control system that has a user programmable memory for storage of instructions to implement specific functions such as: input/output control logic, timing, counting, arithmetic, and data manipulation." Figures 7–1 and 7–2 show modern PLCs.

The PLC is a **factory-hardened** controlling device designed especially for use in a manufacturing setting. Programmable logic controllers can be interfaced with and programmed to control a variety of different types of automated manufacturing equipment and systems. Programmable logic controllers were originally developed to replace hardwired **relay boards** as control devices. The hardwired relay board was an inefficient technological device for controlling manufacturing operations. The boards themselves were large and cumbersome. They required a great deal of time and physical effort to wire properly. When changes or corrections to processes needed to be made, they had to be physically taken apart and rewired. The earliest programmable controllers were thought of as soft wired controllers because they could be programmed and reprogrammed without physically rewiring boards. These early controllers were more efficient than hardwired relay boards, but they served the same purpose.

There are several key words used frequently in any discussion of PLCs:

- input/output (I/O)
- ladder diagram
- safety shutdown

The **input/output (I/O) modules** of a PLC are the key interfaces that allow human beings to interact with the PLC and allow the PLC to interact with automated manufacturing machines and systems. The sophistication of a PLC is, in part, dictated by the number of **connection points** it has on its I/O modules. A low-end PLC will have 256 or fewer points on its I/O modules. A midrange PLC will have from 256 to 1,028 connection points. A top-of-the-line, sophisticated PLC may have as many as 8,000 connection points on its I/O modules. The number of connection points determines the number of inputs the PLC can receive at one time and, correspondingly, the number of outputs it can generate at once. The typical I/O ratio of a PLC is two inputs to one output. In other words, if a PLC receives 100 inputs, it can be expected to generate approximately 50 outputs from those inputs.

FIGURE 7–1 **Programmable logic controllers.** *Photo by David L. Goetsch.*

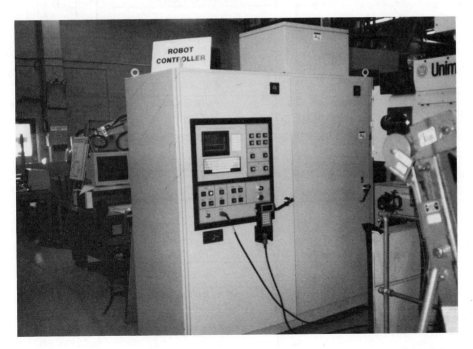

FIGURE 7–2 **Programmable logic controllers for a robot.** *Photo by David L. Goetsch.*

A **ladder diagram** is the type of program used with PLCs designed in the United States. Programmable logic controllers designed in Europe and other countries generally use statement language programs. A ladder diagram is a program in which the logic that will guide the operation of an automated machine is described in a manner similar to rungs on a ladder. There is a limit to the number of rungs on the ladder and a limit to the number of actions that can be described on each rung. Figure 7-3 shows a ladder diagram that is limited to seven rungs and three activities per rung. A PLC interprets this ladder diagram by beginning on the top rung, number 1, and solving for the first activity or function 1. It would then move to rung 2 and solve for function 1. It would then move to rung 3 and solve for function 1 and so on until all of the first functions on each rung had been solved. The controller would then begin again at

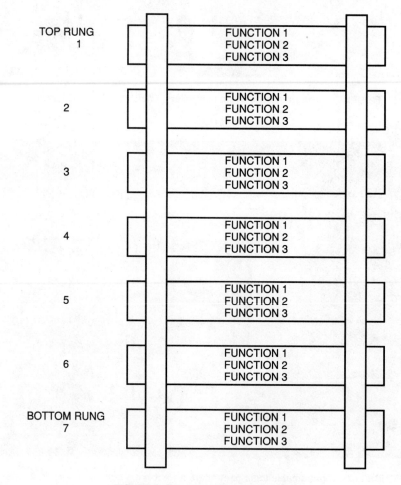

FIGURE 7-3 Ladder diagram concept.

the top rung and solve for function 2 on each rung. It would then begin at the top rung again and solve for function 3. Having interpreted all of the activities on each rung, the PLC would have fully interpreted this particular ladder diagram.

Safety shutdown means shutting down a system in a tightly controlled and specified sequence to ensure that damage does not occur as a result of the shutdown.

PLCs VERSUS PCs

When they were originally developed programmable logic controllers were called programmable controllers (PCs). However, the emergence of the **personal computer** (also **PC**) created confusion. As a result, programmable controllers became programmable logic controllers (PLCs) to avoid confusion. A PLC is a computer. However, it is a special-purpose computer, whereas a PC is a general-purpose computer. Like a PC, a PLC has a **central processing unit (CPU)** with an I/O section. It also has a terminal screen of some sort to display data. Programmable logic controllers also have one of a number of different types of devices that allow human operators to interface with them.

Unlike PCs, PLCs come in specially designed packaging that allows them to withstand the rugged environment of the shop floor. In addition, PLCs must be designed to hold the amount of downtime to an absolute minimum. When a PLC becomes inoperative, an entire manufacturing operation may have to shut down until it is repaired or replaced. For this reason, PLCs are also designed for ease of repair and change-out.

Another difference between PLCs and PCs is that PLCs are programmed using ladder diagrams. Personal computers are programmed using any one of a variety of high-level programming languages. Ladder diagrams are conceptually similar to the schematic instructions that were formerly used to hardwire relay boards. The use of the ladder diagram is tied to this old relay board technology. The rationale is that if a technician could follow the instructions to hardwire a relay board, the same knowledge could be used to develop ladder diagrams. However, ladder diagrams are a cumbersome approach compared with high-level computer languages.

As PLC technology continues to evolve, high-level programming languages are taking the place of ladder diagrams. This allows programs to be written on a PC and then downloaded into a PLC. Any time this can be done, the flexibility of the particular automated system is increased. For example, a manufacturing engineer might develop a program for a PLC using a PC while flying on an airplane or sitting in a motel room. When the engineer arrives back at the manufacturing plant, the program could be downloaded into the PLC and manufacturing operations could begin immediately.

HISTORICAL DEVELOPMENT OF PLCs

The PLC was invented in 1969 by Richard Morley, Director of Advanced Technologies at Gould Electronics Instrumentation and Automation Systems. Prior to this, the hardwired relay board had been the principal method of control other than manual control. General Motors Corporation was the first major manufacturer to use PLCs to replace hardwired relay boards. The operating system for these early controllers was written in assembly language and a memory size of even 4K was considered large. Input and output were accomplished through the use of binary numbers and registers.

As electronics technology evolved, particularly computer-related technology, so did the PLC. The earliest PLCs were only very distant cousins of the computer. However, a top-of-the-line PLC today can almost be considered the twin of the modern computer. In fact, regular general-purpose PCs and related products are beginning to be used in place of PLCs. As a result, producers of PLCs are beginning to design in new and expanded capabilities.

In an article in *Manufacturing Engineering*, Robert Eisenbrown, Allen-Bradley Company, listed some of the capabilities that the PLC of the future would have. According to Eisenbrown, the modern PLC has all of the following traditional capabilities:

1. a control-oriented programming language
2. high-speed, consistent user program execution cycles
3. dedicated processing for consistent I/O updating
4. continuous error checking and fault monitoring
5. immediate, orderly shutdown with last-state retention in the event of workstation error or failure
6. a shop-toughened package

According to Eisenbrown, in addition to these traditional capabilities, the PLC of the future will also have to have additional new capabilities (Figure 7–4). Among the new capabilities of the PLC of the future are the following:

1. the ability to accommodate the integration of automated machines and systems

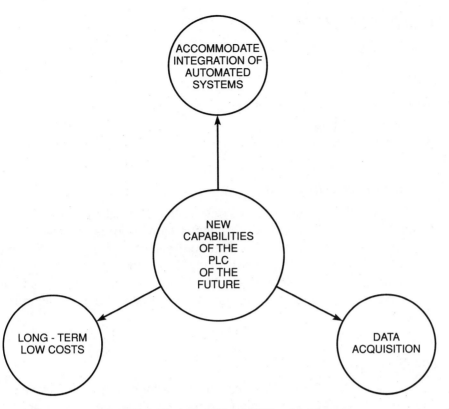

FIGURE 7-4 New capabilities of PLCs of the future.

2. data acquisition
3. long-term low costs

     The PLC was originally developed to allow for the control of individual automated machines. No provision was made in the design of PLCs to allow for **networking** or communication among individual islands of automation. However, with integration clearly representing the future of manufacturing, modern PLCs will have to have integration capabilities. Data acquisition is another capability the modern PLC will have to have if it is to remain an integral part of the modern manufacturing plant. Finally, the PLC of the future will have to be the lowest cost alternative in the area of controllers on a long-term basis. This will not be as easy to accomplish as with earlier models that did not require such sophisticated capabilities. This is why many leaders in the field of manufacturing automation and integration feel the PLC of the future will more closely resemble a factory-hardened PC than the traditional PLC.

CONFIGURATION OF THE PLC

The overall configuration of a PLC consists of four principal components (Figure 7–5):

1. central processing unit (CPU)
2. input/output (I/O) modules
3. programming panel
4. power supply

Central Processing Unit

As with any computerlike device, the CPU is the key component of the PLC. The CPU of a PLC serves the same purpose as the CPU in a computer. It is the component that controls the overall PLC as a unit, integrates the various components of the PLC, and performs all logic and arithmetic problem-solving tasks. Most communication between the PLC and other machines or systems being controlled is accomplished by the CPU.

I/O Modules

A PLC's I/O modules allow it to interface with its environment. The modules contain connection points that allow the PLC to accept

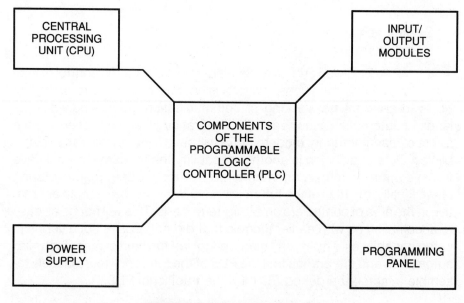

FIGURE 7–5 Components of the PLC.

input from the outside and generate output that it sends to the machines and systems it controls. A low-end PLC will have 256 or fewer connection points. A high-end PLC might have as many as 8,000 connection points.

Programming Panel

The **programming panel** is the principal means for interaction between human operators and the PLC. Historically, the programming panel has been the principal means for entering a program into the PLC. However, modern PLCs can accept programs downloaded from a PC. More and more, the modern PLC uses a PC in the place of the programming panel. This is a trend that will continue until the programming panel is completely replaced by the PC.

Power Supply

The **power supply** is a key component in a PLC configuration. Power interruptions or even fluctuations in power can cause a PLC to malfunction. This is the "Achilles heel" of the PLC since power interruptions and fluctuations are common in a manufacturing setting. Because of this, modern PLCs are being designed with power storage capabilities to ensure a steady flow of power in the event of interruptions or fluctuations. At the least, the storage capability of the modern PLC will allow the controller to shut down the systems it controls in an organized and orderly manner so that neither the product nor the machines being controlled are damaged as a result of a power interruption.

OPERATION OF THE PLC

When a PLC operates, it processes a program that has five components. These five components represent steps in the operation of the PLC (Figure 7–6):

1. process the self test program
2. write to the I/O
3. process the logic-solving program
4. read the I/O
5. handle communications

The PLC continually goes through this five-step process.

Processing the Self-Test

Each time a PLC is turned on, it undertakes a number of self-tests to ensure that all components are operating properly. As the PLC

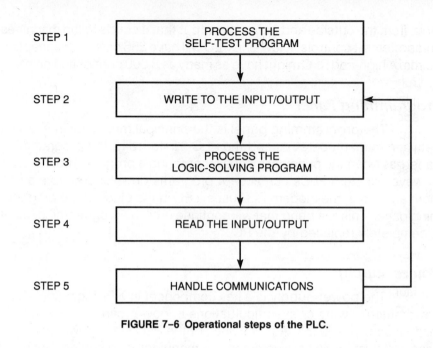

STEP 1	PROCESS THE SELF-TEST PROGRAM
STEP 2	WRITE TO THE INPUT/OUTPUT
STEP 3	PROCESS THE LOGIC-SOLVING PROGRAM
STEP 4	READ THE INPUT/OUTPUT
STEP 5	HANDLE COMMUNICATIONS

FIGURE 7–6 Operational steps of the PLC.

proceeds through its various self-tests, it shuts down if it discovers a malfunction to ensure that the system is not damaged. If all the self-tests are satisfactory, the PLC proceeds to step 2. Once the self-test step has been accomplished, the PLC uses a built-in **watch-dog timer** as a guard against system failure. The watchdog timer monitors a signal sent from the CPU. As long as the signal is received, the system is operating. Should the system malfunction, the signal would stop and the PLC would know immediately that a problem exists. The action the PLC takes on losing the signal depends on the way it has been programmed.

Write to the I/O

This step involves transferring I/O data housed in the output buffer in the CPU to the I/O modules. These data are processed, thereby bringing all I/O points into the proper and corresponding state. The PLC always undertakes this step before attempting to read to be sure it is starting at a known state.

Process the Logic-Solving Program

This is the step that causes people to associate with the operation of a PLC with a ladder. Logic solving involves solving the ladder diagram. The PLC accomplishes this by starting at the top rung of the ladder and solving function 1 on that rung, then moving to function 1 on rung 2. From there it moves to function 1 on rung 3 and from there to function 1 on rung 4 and so on, until all first functions on all rungs of the ladder diagram have been solved. At this point it begins again at the top rung of the ladder, solves function 2 at that rung, function 2 on the next rung, and continues this process, working its way down the ladder until all second functions have been solved. This process continues until the PLC has processed the entire logic solving program.

Read the I/O

Once the PLC has processed the entire logic-solving program, it undertakes the fourth step, which involves reading the I/O. Information going to or coming from the I/O modules is checked and double-checked to ensure the proper interpretation at this point. Once this step has been accomplished the PLC moves to the final step.

Handle Communications

This step involves checking all external communication interfaces such as a computer, programming panel, or any other type of human communications interface. At this point, the PLC can detect any input coming from external sources.

The PLC has now accomplished one complete scan. A **scan** is the process of going through each step outlined in Figure 7–6. Note in Figure 7–6 that it is not necessary for the PLC to process step 1, the self-test program, except when it is initially powered up. After that point, the continuous loop is steps 2 through 5, so long as the built-in watchdog timer continues to generate the appropriate signal. Figure 7–6 can be altered even further to produce a faster scan by eliminating step 5 or by undertaking step 5 only periodically.

APPLICATIONS OF PLCs

Programmable logic controllers are used for controlling machines and equipment in a number of different industrial settings. Programmable logic controllers are used outside of manufacturing for controlling elevators, irrigation systems, and traffic lights. In manufacturing, PLCs are used to control machine tools, robots, automated assembly systems, materials handling systems, flexible manufacturing cells, and numerous other machine and systems.

Figure 7–7 shows a series of PLCs that control several different systems. Note the PCs in the foreground. Figure 7–8 is a PLC used to control an industrial robot. Figure 7–9 shows retrofitted PCs being used

FIGURE 7–7 Programmable logic controllers with PCs in the foreground. *Photo by David L. Goetsch.*

FIGURE 7–8 Programmable logic controller for controlling an industrial robot. *Photo by David L. Goetsch.*

as controllers for a flexible manufacturing workstation. Figure 7–10 is a computer-controlled manufacturing system. The PLC used to control the system can be seen on the right.

FIGURE 7–9 Retrofitted PCs being used as PLCs. *Photo by David L. Goetsch.*

FIGURE 7–10 Manufacturing system controlled by a PLC. *Courtesy of Cincinnati Milacon.* Safety equipment may have been removed or opened to clearly illustrate the product and must be in place prior to operation.

As technology continues to evolve, the PC will be used more frequently as a platform for PLCs. By adding factory-hardened packaging, an interface to discrete I/O points, a ladder diagram compiler, a display driver, and control software to the modern PC, manufacturing personnel can convert it to a PLC. Such retrofitting will be done more and more in the future. Figure 7–11 shows a PC that has been retrofitted for use as a PLC for controlling an industrial robot system. Note the robot's teach pendant on the table beside the PC/PLC.

FIGURE 7–11 Retrofitted PC being used as a PLC. *Photo by David L. Goetsch.*

KEY TERMS

Programmable logic
      controller (PLC)
Factory-hardened
Relay boards
Input/output (I/O) modules
Connection points
Ladder diagram
Safety shutdown

Personal computer (PC)
Central processing
      unit (CPU)
Networking
Programming panels
Power supply
Watchdog timer
Scan

Chapter Seven REVIEW

1. What is a programmable logic controller?
2. Why were PLCs developed?
3. Explain the following terms:
    Soft wired
    I/O
    Ladder diagram
    Safety shutdown
4. How many connection points might a sophisticated PLC have on its I/O modules?
5. Explain the difference between a PLC and a PC.
6. List at least five traditional capabilities of a PLC.
7. List three capabilities of future PLCs.
8. List and explain the four main components of a PLC.
9. Explain the achilles heel of PLCs.
10. List and explain the five steps in operating a PLC.
11. What is a watchdog timer?
12. Explain the term scan.
13. What is meant by networking?

These case studies were provided by the Society of Manufacturing Engineers (SME). They are excerpted from the SME's *Manufacturing Insights*® series of videotapes. The case studies contained herein give students actual examples of the way the advanced manufacturing technologies covered in this chapter are being applied in "real-world" manufacturing settings.

As you read each case study, relate the examples cited to the material presented in the text of the chapter. This combination of textbook information and real-world examples will be particularly valuable to your understanding of advanced manufacturing technologies.

MOLEX CORPORATION

Programmable logic controllers have proven highly reliable and effective in increasing productivity and greatly reducing setup time at the Molex Corporation assembly plant in Lincoln, Nebraska.

Allen West, production manager, explains:

> At the Lincoln assembly plant for Molex, we make electrical connectors in a variety of sizes from 2 circuit through 24 circuit for a number of different applications that go into computers, automobiles, vending machines, and stereos. We make everything for commercial products.
>
> The setup for our original equipment was all manually done as far as making physical adjustments to the equipment, changing pins and toggle switches, and adjusting physical mechanical operations to the machine. The setup of the machine took approximately 30 minutes and required one person's full attention for that time.
>
> When we went to the programmable controller, we were able to program into the machine exactly what we wanted the machine to do, without changing any of the mechanical variation of the machine. It was strictly electronic. The setup time went down to approximately 10 minutes per change and we do probably in excess of 300 setups in a month's time. The saving is tremendous for us.

At Molex, many of the parts are very small and they are produced in great quantity, so precision and reliability of the equipment is a high priority. On the assembly machines, the plastic connector bodies are loaded into a track, and the metal pins to be staked into the bodies are fed continuously from a large roll. Proximity sensors along the track verify that the bodies are present just before they enter the assembly station, where the pins are inserted into the bodies.

Because of the critical nature of the applications for the parts and the wide variety of combinations of pins present and pins absent, 100% inspection is required of all connectors. Fewer than 0.5% of all parts are rejected as defective, or less than 500 per million.

All of the actions of the machine are controlled by the programmable controller, which also counts the number of parts produced and automatically shuts down the machine when the preset production number has been reached.

Because of the great number of configurations in the connectors and the wide variety of quantities produced to meet varying customer requirements, frequent reprogramming of the controllers is necessary. This is done quickly and easily with a handheld programmer, usually without having to make any physical changes to the machinery.

Allen West continues:

> Since the change in the machine and putting the programmable controller on the machine, we were able to double our output per hour by utilizing the capacity of the controller. In the future, we plan to look at converting or incorporating into the controllers a networking system to put into our office and into our computers—a way to tell the machines what to do off-line; to tell the machines when to shut down after running a 600,000-piece order, which for us is normal; and to tell the machine what the next order will be, which will allow us to have people on the floor only to take the components to the machines. The machine will already be setup for what we're going to run next.
>
> The programmable controllers have reduced the setup time for my lead man and supervisors. It has cut the setup time in half. The reliability of the programmable controllers has proven to be very efficient and the capacities—capabilities of what we're going to use them for in the future is endless as far as we can see.

(continued)

BABCOCK & WILCOX

Programmable controllers were responsible for a substantial increase in productivity and quality performance at the Babcock & Wilcox Nuclear Equipment Division plant in Barberton, Ohio. The nuclear equipment products produced in this plant include steam generators, pressurizers, closure heads, and other components for nuclear power generation facilities. The volume of individual products is low and quality is a top priority.

At this facility, programmable controllers are used primarily for controlling multiaxis welding manipulators for seam welding and bore cladding prior to machining.

Hot and cold wire sub-arc and TIG welding are performed at the plant, both of which are monitored by the programmable controllers. These units control all feeds and axis positioning, as well as the weld voltage and current that are vital to producing the best quality welds.

One major operation is the use of welders to clad the inside of large bores for machining. A programmable controller is used to establish the center of the bore as a reference for guiding the welder and to control the movements of the welder inside the bore, as well as the current, voltage and wire feed rate. A programmable controller is also used to control an automatic drilling operation, permitting the selective positioning of holes in the workpiece to meet the requirements of individual parts. This system replaced a previous hand setup method that was cumbersome, time consuming, and less accurate.

In addition to the welding and drilling operations, programmable controllers are used at the Barberton plant to control air and water at the powerhouse, the 3000-psi water system for the pierce and draw operation, and several other special machines.

The use of programmable controllers in the welding operation at the Babcock & Wilcox Nuclear Equipment plant has increased production, shortened setup time, and reduced both weld defects and maintenance downtime. They have also provided greater flexibility in changing welding configurations to meet changes in customer specifications.

Chapter Eight

Major Topics Covered

- Flexible Manufacturing System Defined
- Overview of Flexible Manufacturing
- Historical Development of Flexible Manufacturing
- Rationale for Flexible Manufacturing
- Flexible Manufacturing System Components
- Flexible Manufacturing Cells
- Artificial Intelligence and Flexible Manufacturing
- Machining Centers
- *Case Study: Agnew Machine Company*
- *Case Study: Yamazaki Mazak*

Automated Assembly

Automated Guided Vehicles

CAD/CAM

Numerical Control

Industrial Robots

Lasers in Manufacturing

Programmable Logic Controllers

Flexible Manufacturing

Computer Integrated Manufacturing

Other Related Technologies

FLEXIBLE MANUFACTURING SYSTEM DEFINED

The evolution of manufacturing can be represented graphically as a continuum as shown in Figure 8–1. As this figure shows, manufacturing processes and systems are in a state of transition from manual operation to the eventual realization of fully integrated manufacturing. The step preceding computer-integrated manufacturing is called **flexible manufacturing**.

Flexibility is an important characteristic in the modern manufacturing setting. It means that a manufacturing system is versatile and adaptable, while also capable of handling relatively high production runs. A flexible manufacturing system is versatile in that it can produce a variety of parts. It is adaptable in that it can be quickly modified to produce a completely different line of parts. This flexibility can be the difference between success and failure in a competitive international marketplace.

It is a matter of balance. Stand-alone computer numerical control (CNC) machines have a high degree of flexibility, but are capable of relatively low-volume production runs. At the opposite end of the spectrum, transfer lines are capable of high-volume runs, but they are not very flexible. Flexible manufacturing is an attempt to use technology in such a way as to achieve the optimum balance between flexibility and production runs. These technologies include automated materials, handling, group technology, and computer and distributed numerical control.

A **flexible manufacturing system (FMS)** is an individual machine or group of machines served by an automated materials handling system that is computer controlled and has a tool handling capability. Because of its tool handling capability and computer control, such a system can be continually reconfigured to manufacture a wide variety of parts. This is why it is called a flexible manufacturing system.

The key elements necessary for a manufacturing system to qualify as an FMS are as follows (Figure 8–2):

1. computer control
2. automated materials handling capability
3. tool handling capability

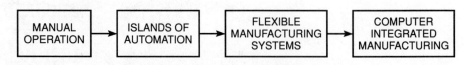

FIGURE 8–1 The evolution of manufacturing.

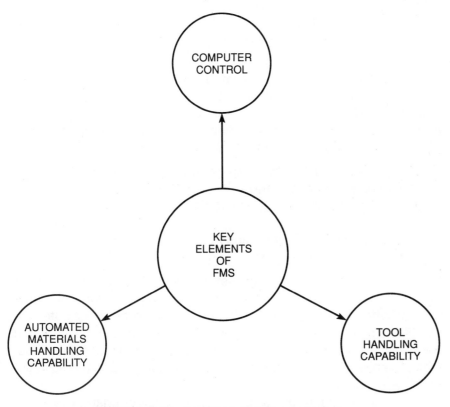

FIGURE 8–2 Key elements in flexible manufacturing.

Flexible manufacturing represents a major step toward the goal of fully integrated manufacturing in that it involves integration of automated production processes. In flexible manufacturing, the automated manufacturing machine (i.e., lathe, mill, drill) and the automated materials handling system share instantaneous communication via a computer network. This is integration on a small scale.

OVERVIEW OF FLEXIBLE MANUFACTURING

Flexible manufacturing takes a major step toward the goal of fully integrated manufacturing by integrating several automated manufacturing concepts:

1. **computer numerical control (CNC)** of individual machine tools
2. **distributed numerical control (DNC)** of manufacturing systems

3. **automated materials handing** systems
4. **group technology** (families of parts)

When these automated processes, machines, and concepts are brought together in one integrated system, an FMS is the result. Humans and computers play major roles in an FMS. The amount of human labor is much less than with a manually operated manufacturing system, of course. However, humans still play a vital role in the operation of an FMS. Human tasks include the following:

1. equipment troubleshooting, maintenance, and repair
2. tool changing and setup
3. loading and unloading the system
4. data input
5. changing of parts programs
6. development of programs

Flexible manufacturing system equipment, like all manufacturing equipment, must be monitored for "bugs," malfunctions, and breakdowns. When a problem is discovered, a human troubleshooter must identify its source and prescribe corrective measures. Humans also undertake the prescribed measures or repair the malfunctioning equipment. Even when all systems are properly functioning, periodic maintenance is necessary.

Human operators also set up machines, change tools, and reconfigure systems as necessary. The tool handling capability of an FMS decreases, but does not eliminate, human involvement in tool changing and setup. The same is true of loading and unloading the FMS. Once raw material has been loaded onto the automated materials handling system, it is moved through the system in the prescribed manner. However, the original loading onto the materials handling system is still usually done by human operators, as is the unloading of finished products.

Humans are also needed for interaction with the computer. Humans develop parts programs that control the FMS via computers. They also change the programs as necessary when reconfiguring the FMS to produce another type of part or parts. Humans also input data needed by the FMS during manufacturing operations. Humans play less labor-intensive roles in an FMS, but the roles are still critical.

Control at all levels in an FMS is provided by computers. Individual machine tools within an FMS are controlled by CNC. The overall system is controlled by DNC. The automated materials handling system is computer controlled, as are other functions including data collection, system monitoring, tool control, and traffic control. Human/computer interaction is the key to the flexibility of an FMS.

HISTORICAL DEVELOPMENT OF FLEXIBLE MANUFACTURING

Flexible manufacturing was born in the mid-1960s when the British firm Molins, Ltd. developed its **System 24**. System 24 was a real FMS. However, it was doomed from the outset because automation, integration, and computer control technology had not yet been developed to the point where they could properly support the system. This first FMS was a development that was ahead of its time. As such, it was eventually discarded as unworkable.

Flexible manufacturing remained an academic concept through the remainder of the 1960s and 1970s. However, with the emergence of sophisticated computer control technology in the late 1970s and early 1980s, flexible manufacturing became a viable concept. The American pioneer of flexible manufacturing was Kearney & Trecker, still a leading supplier. Other leading producers of FMSs are Cincinnati Milacron, Giddings & Lewis, and White Sunstrand (Figure 8–3).

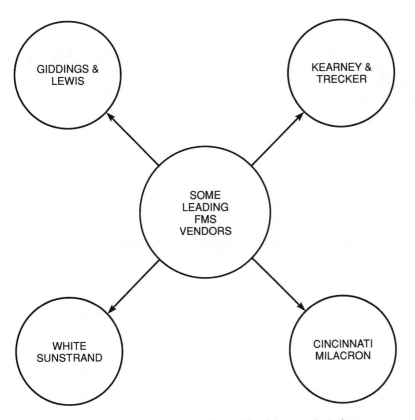

FIGURE 8–3 Some leading suppliers of flexible manufacturing.

The principal users of flexible manufacturing in the United States have been manufacturers of automobiles, trucks, and tractors. In Japan, the principal user in the machine tool industry. In 1985, there were approximately 200 FMS installations in the world. Of these, approximately 50 were located in the United States. There are now approximately 250 FMS installations in the world with about half of them in Europe, one-fourth in the United States, and the remaining one-fourth in Japan.

RATIONALE FOR FLEXIBLE MANUFACTURING

In manufacturing there have always been trade-offs between production rates and flexibility. At one end of the spectrum are transfer lines capable of high production rates, but low flexibility. At the other end of the spectrum are independent CNC machines that offer maximum flexibility, but are only capable of low production rates (Figure 8–4). Note from the figure that flexible manufacturing falls in the middle of the continuum. There has always been a need in manufacturing for a system that could produce higher volume and production runs than could independent machines while still maintaining flexibility.

Transfer lines are capable of producing large volumes of parts at high production rates. The line takes a great deal of setup, but can turn out identical parts in large quantities. Its chief shortcoming is that even minor design changes in a part can cause the entire line to be shut down and reconfigured. This is a critical weakness because it means that transfer lines cannot produce different parts, even parts from within the same family, without costly and time-consuming shutdown and reconfiguration.

Traditionally, CNC machines have been used to produce small volumes of parts that differ slightly in design. Such machines are ideal for this purpose because they can be quickly reprogrammed to accommodate minor or even major design changes. However, as independent

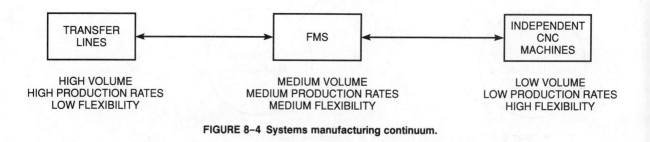

FIGURE 8–4 Systems manufacturing continuum.

machines, they cannot produce parts in large volumes or at high production rates.

An FMS can handle higher volumes and production rates than independent CNC machines. They cannot quite match such machines for flexibility, but they come close. What is particularly significant about the middle ground capabilities of flexible manufacturing is that most manufacturing situations require medium production rates to produce medium volumes with enough flexibility to quickly reconfigure to produce another part or product. Flexible manufacturing fills this long-standing void in manufacturing.

Flexible manufacturing, with its middle ground capabilities, offers a number of advantages for manufacturers:

1. flexibility within a family of parts
2. random feeding of parts
3. simultaneous production of different parts
4. decreased setup times/lead time
5. efficient machine usage
6. decreased direct and indirect labor costs
7. ability to handle different materials
8. ability to continue some production if one machine breaks down

Flexibility Within a Family of Parts

A **family of parts** is a group of parts that is similar enough in design to require similar production processes. An FMS is flexible enough that it can be quickly reconfigured to produce a wide variety of parts so long as they fall within the same family of parts. Figure 8–5 shows an example of two different parts within the same family of parts.

Random Feeding of Parts

An FMS can be set up and programmed to accommodate random feeding of parts within the family being produced. This involves ensuring that the necessary tooling is set up on the appropriate machines and the necessary controls are programmed in. With this done, parts may be introduced randomly by identifying each part to the controller as it is introduced into the system. The controller then routes the part to the appropriate machines in the processes within the FMS.

Simultaneous Production

In addition to its random feeding capability, an FMS also has a **simultaneous production** capability. This means that different parts

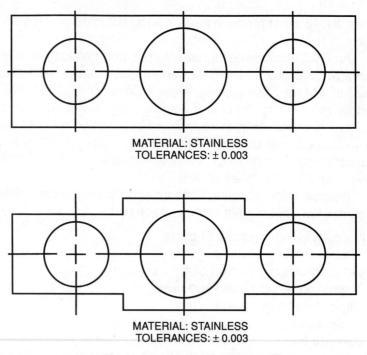

MATERIAL: STAINLESS
TOLERANCES: ± 0.003

MATERIAL: STAINLESS
TOLERANCES: ± 0.003

FIGURE 8–5 Parts within a family of parts.

within a family of parts can be processed at the same time as the rest of the system. Again, this is due to the flexibility that comes from programmed operations. The key lies in effective production planning and programming.

Decreased Setup and Lead Time

An important advantage of flexible manufacturing is that it requires less setup time. This, in turn, means that an FMS requires less lead time. Setting up machines for manufacturing operations involves

1. outfitting them with the proper tools
2. placing raw materials into the fixtures and making the necessary adjustments

Each time a setup operation is required, a traditional manufacturing system must be shut down. Tools must be retrieved from the tool crib, raw materials must be retrieved from storage, and machines must be reconfigured. Because of this, extensive lead time is often necessary.

With flexible manufacturing, tooling and raw stock can be set up off line. In this way, changing tools is just a matter of placing preset tools on the appropriate machines. An individual machine might be loaded with 50 or more different tools at once. Then, rather than changing tools, the system is simply programmed as to when it should use each tool. Hence, breakdown and resetup time are saved. Raw material is also provided off line, usually on pallets. The pallets are configured for the specific operations that will take place. The pallets contain all of the necessary fixtures. Again, this saves on breakdown and setup time.

More Efficient Machine Usage

The combined advantages of flexible manufacturing lead to another advantage: more efficient machine usage. In traditional manufacturing systems, individual machines are set up to perform specific operations. If a given part requires three different processes, two machines in a system are typically idle, waiting for processing on the other machines. Only the machine processing the current operation is normally in use. This results in inefficient usage.

Flexible manufacturing systems solve this problem. If that same part was produced on an FMS and 100 copies were to be produced, the system could be set up and programmed to produce the part simultaneously. While station 1 performs its processes on a group of workpieces, stations 2 and 3 would be performing all or some of their processes on other groups of workpieces. The workpieces then rotate through each station as needed. The result is more efficient use of machines.

Decreased Direct and Indirect Labor Costs

In traditional manufacturing systems and even with individual CNC machines, there is typically one human operator for each machine. Add to this the human labor involved in materials handling away from the machine and additional manufacturing systems, including CNC machines. The amount of on-line and off-line labor decreases with flexible manufacturing. This can be attributed to the following:

1. automated as opposed to human materials handling with flexible manufacturing
2. automated as opposed to human control
3. off-line as opposed to on-line setup and tooling preparation

FLEXIBLE MANUFACTURING SYSTEM COMPONENTS

An FMS has four major components:

1. machine tools
2. control system
3. materials handling system
4. human operators

Machine Tools

A flexible manufacturing system uses the same types of machine tools as any other manufacturing system, be it automated or manually operated. These include lathes, mills, drills, saws, and so on. The type of machine tools actually included in an FMS depends on the setting in which the system will be used. Some FMSs are designed to meet a specific, well-defined need. In these cases, the machine tools included in the system will be only those necessary for the planned operations. Such a system would be known as a dedicated system.

In a job shop setting, or any other setting in which the actual application is not known ahead of time or must necessarily include a wide range of possibilities, machines capable of performing at least the standard manufacturing operations would be included. Such systems are known as general-purpose systems. Figures 8–6 and 8–7 are examples of typical FMS configurations.

Control Systems

The **control system** for an FMS serves a number of different control functions for the system:

1. storage and distribution of part programs
2. work flow control and monitoring
3. production control
4. system/tool control/monitoring

The control system of an FMS accepts, stores, and distributes parts programs. These are the CNC programs that guide the operation of individual machines and workstations within the system in performing the turning, cutting, drilling, and other processes necessary to produce parts.

Regulating the flow of workpieces from station to station for both primary and secondary materials handling systems and monitoring the locations of workpieces within the system are important control tasks. Different parts must be directed to different workstations in the most efficient order. Different parts require different speeds and feed rates. These types of production-oriented controls represent another important control function in FMS.

The overall system must be monitored and controlled as must individual tools within it. The control system must continually collect and store data that can periodically be output in the form of performance reports. The wear on individual tools should be monitored continually so that worn tools can be changed as needed and reports on the projected

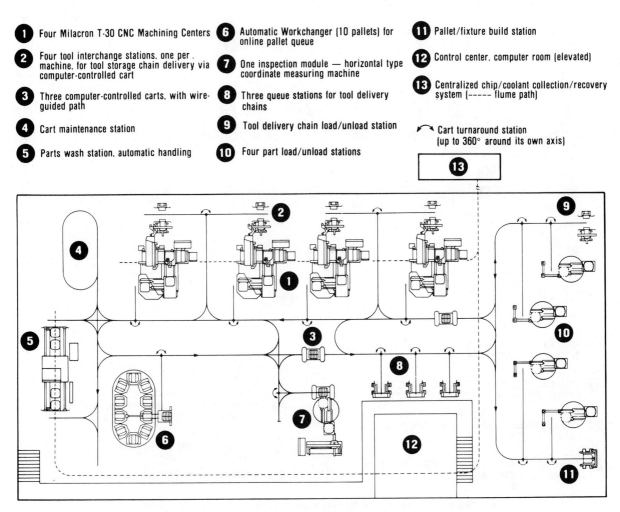

1 Four Milacron T-30 CNC Machining Centers

2 Four tool interchange stations, one per machine, for tool storage chain delivery via computer-controlled cart

3 Three computer-controlled carts, with wire-guided path

4 Cart maintenance station

5 Parts wash station, automatic handling

6 Automatic Workchanger (10 pallets) for online pallet queue

7 One inspection module — horizontal type coordinate measuring machine

8 Three queue stations for tool delivery chains

9 Tool delivery chain load/unload station

10 Four part load/unload stations

11 Pallet/fixture build station

12 Control center, computer room (elevated)

13 Centralized chip/coolant collection/recovery system (----- flume path)

⌒ Cart turnaround station (up to 360° around its own axis)

FIGURE 8–6 Flexible manufacturing system. *Courtesy of Cincinnati Milacron.*

versus actual lives of tools can be produced. These are important functions of the control system.

Materials Handling System

The automated materials handling system is a fundamental component that helps mold a group of independent CNC machines into a comprehensive FMS. The system must be capable of accepting workpieces mounted on pallets and moving them from workstation to workstation as needed. It must also be able to place workpieces "on hold" as they wait to be processed at a given workstation.

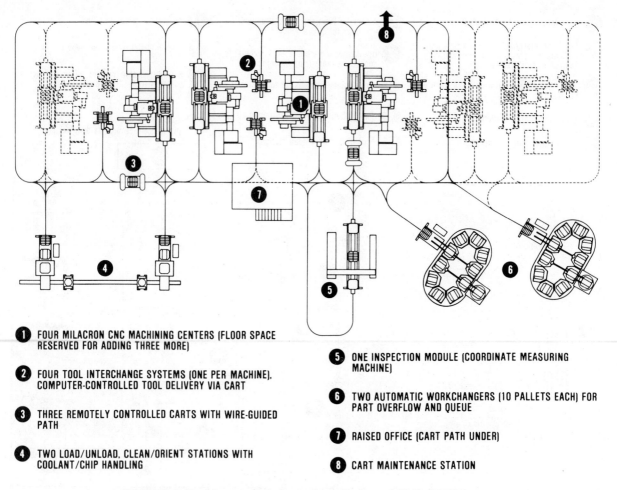

① FOUR MILACRON CNC MACHINING CENTERS (FLOOR SPACE RESERVED FOR ADDING THREE MORE)

② FOUR TOOL INTERCHANGE SYSTEMS (ONE PER MACHINE). COMPUTER-CONTROLLED TOOL DELIVERY VIA CART

③ THREE REMOTELY CONTROLLED CARTS WITH WIRE-GUIDED PATH

④ TWO LOAD/UNLOAD, CLEAN/ORIENT STATIONS WITH COOLANT/CHIP HANDLING

⑤ ONE INSPECTION MODULE (COORDINATE MEASURING MACHINE)

⑥ TWO AUTOMATIC WORKCHANGERS (10 PALLETS EACH) FOR PART OVERFLOW AND QUEUE

⑦ RAISED OFFICE (CART PATH UNDER)

⑧ CART MAINTENANCE STATION

FIGURE 8–7 Flexible manufacturing system. *Courtesy of Cincinnati Milacron.*

The materials handling system must be able to unload a workpiece at one station and load another for transport to the next station. It must accommodate computer control and be completely compatible in that regard with other components in the flexible manufacturing system. Finally, the materials handling system for an FMS must be able to withstand the rigors of a shop environment.

Some FMSs are configured with automated guided vehicles (AGVs) as a principal means of materials handling. Figure 8–8 is an example of such a system. This system has ten main components. Component 2 is an AGV used to move material from station to station. Figures 8–6 and 8–7 are also FMSs that use AGVs for materials handling.

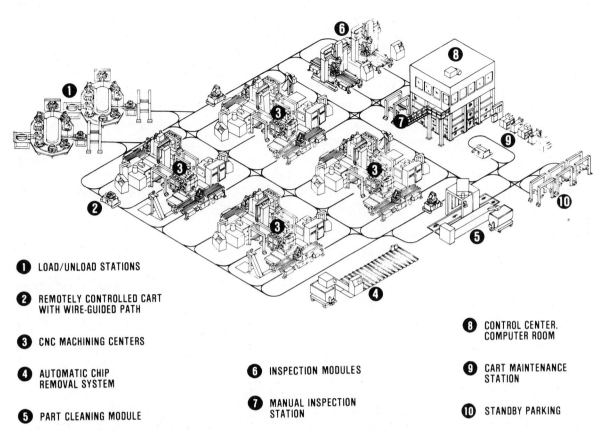

1　LOAD/UNLOAD STATIONS

2　REMOTELY CONTROLLED CART
　　WITH WIRE-GUIDED PATH

3　CNC MACHINING CENTERS

4　AUTOMATIC CHIP
　　REMOVAL SYSTEM

5　PART CLEANING MODULE

6　INSPECTION MODULES

7　MANUAL INSPECTION
　　STATION

8　CONTROL CENTER,
　　COMPUTER ROOM

9　CART MAINTENANCE
　　STATION

10　STANDBY PARKING

FIGURE 8–8 Flexible manufacturing system with an AGV for materials handling. *Courtesy of Cincinnati Milacron.*

Human Operators

The final component in an FMS is the human component. Although flexible manufacturing as a concept decreases the amount of human involvement in manufacturing, it does not eliminate it completely. Further, the roles humans play in flexible manufacturing are critical roles.

FLEXIBLE MANUFACTURING CELLS

Flexible manufacturing cells (FMCs) are dedicated groups of workstations within a larger FMS. Such cells are segregated components of an FMS. The reasons for segregating a cell within a system are varied:

1. working with dangerous or hazardous materials
2. noise
3. different operator tasks required
4. different materials required

The development of FMCs followed that of FMSs, although it might have worked better had the order been reversed. Flexible manufacturing cells are actually small FMSs. As such, they are less expensive to develop and implement and more likely to succeed. For this reason, FMCs are growing in use more rapidly than the larger FMSs.

The FMC is actually the FMS concept in microcosm. As such, it offers several advantages over FMSs:

1. less initial capital outlay
2. less sophisticated computer control
3. easy to learn to operate
4. more easily understood by management and production personnel

In its simplest form, an FMC is a group of related CNC machines with common computer control and a common materials handling system.

ARTIFICIAL INTELLIGENCE AND FLEXIBLE MANUFACTURING

A technology that is having a positive impact on flexible manufacturing is **artificial intelligence**. A difficult problem to overcome in flexible manufacturing is the inability of automated machines to mimic such basic human capabilities as adjusting appropriately to differences in the size, shape, or orientations of objects. This is particularly true in assembly processes. Artificial intelligence is helping to solve this dilemma.

For example, an assembly worker whose job is to retrieve small parts from a feeder bin and insert them into the appropriate holes in a plate, has many human attributes to assist him. These include sight, hand-eye coordination, reasoning abilities, logic, judgment, and experience. If the assembly worker picks up a part that is not properly oriented or if on his first attempt to insert the part, it does not properly seat, these human capabilities allow him to adjust appropriately. Even the most sophisticated automated assembly systems used in flexible manufacturing settings cannot completely mimic these capabilities. Artificial intelligence is an attempt to increase the number of human characteristics computers and computer-controlled systems can mimic.

Artificial Intelligence Defined

Artificial intelligence is the ability of a computer to imitate human intelligence and, thereby, make intelligent decisions. Computer-controlled systems that apply artificial intelligence to everyday settings are called expert systems.

In any discussion of artificial intelligence, several key words and phrases are frequently used:

Algorithm: A special computer program that will solve selected problems within a given time frame.

Early Vision: Computer calculations that allow systems to see by providing low-level data such as spatial and geometrical information.

Higher Level Vision: Computer calculations that allow systems to accomplish higher level tasks such as smart movement within an environment, object recognition, and reasoning about objects.

Knowledge Engineering: A process through which knowledge is collected from experts in a given field and converted into a computable format.

Neurocomputering: An approach to performing mathematical calculations on a computer that is based on the way the human nervous system operates.

Humans can make logical, reasoned adjustments in a work setting because they can quickly collect information, access it against the sum total of their human experience, and evaluate known relationships among various items of information. The science of artificial intelligence attempts to imitate this process with computers.

Humans attempt to create an experience base in computers by feeding them all known information about a given subject. This information is then used by the computer in making decisions. This is why the concept is called artificial intelligence. The computer does not really think. Rather, it simply searches its memory for the appropriate information. If the information is there, the computer uses it in making logical decisions. The key is in feeding the computer enough relevant information.

Historical Development of Artificial Intelligence

In the 1950s, scientists interested in artificial intelligence moved computers from number processors to symbol processors. This was a major step forward. The next step involved the development and use of algorithms for solving problems.

The 1960s saw the development of heuristic search. This cut down significantly on the amount of space through which a computer would search for solutions. Heuristics are rules of thumb that limit the size of the space searched by getting the computer in the ballpark from the outset. The one weakness of the heuristic search is that, although it locates a solution faster, the solution is not necessarily the best solution. However, a good heuristic search gives up only a little in finding a solution, while gaining a great deal in limiting the size of the search area.

The 1970s were characterized by the development of expert systems. This development resulted from the realization of scientists that intelligence would not be achieved by searching through general space. They began to see the need for feeding task specific information to the computer that would give it an experience base relating to a specific task domain. The result was the development of expert systems.

Expert systems have two components:

1. an inference engine
2. an experience or knowledge base

The first component controls the application of information contained in the second component. The development and use of expert systems raised questions about the way knowledge is represented.

Improving the ways knowledge is represented so that it is more explicit but, at the same time, more concise characterized the 1980s. The 1990s will see a rapid growth in the use of artificial intelligence in automated assembly and other manufacturing applications as the concept of machine learning evolves.

It is generally accepted that a key characteristic of human intelligence is the ability to learn. For artificial intelligence to reach its potential, the concept of machine learning must be fully developed. Machines learn in one of three ways:

1. parameter adjustment
2. concept formation
3. evolution of structure

Parameter adjustment is the most basic of the three approaches to machine learning. It involves adjusting the values of the parameters of a predetermined representation. It can affect only the values and not the structure of the predetermined representation. Concept formation involves grouping related objects into categories or groups. Evolution of structure makes use of neurocomputing. This involves the parallel activity of elements that are able to communicate the results of computations among themselves.

Artificial Intelligence in Flexible Manufacturing

Artificial intelligence has applications in a number of manufacturing settings such as robotics, automated materials handling, and automated assembly. Flexible manufacturing brings these areas together in one cell or system. In such a setting, artificial intelligence can be used to allow robots and other systems to duplicate such human capabilities as vision and language processing. It can help improve the assembly skills of robots, the materials handling skills of other machines, and the ability of information management systems used in flexible manufacturing settings.

MACHINING CENTERS

One step down on the manufacturing ladder from the FMC is the **machining center**. A machining center is a multipurpose CNC machine tool that has an automatic tool changing capability. Two characteristics distinguish a machining center from a stand-alone CNC machine:

1. multipurpose configuration
2. automatic tool changing capability

Figure 8–9 shows several examples of machining centers. Note the similarities among these different configurations. Although three- to five-axis machine centers are the most widely used, centers are available with as many as nine axes. Machining centers can be operated alone, as part of an FMC, or as part of an FMS. They offer a number of advantages, the most important of which is versatility. A typical machining center can perform such functions as milling, drilling, boring, reaming, contouring, and threading. This versatility leads to the other advantages such as decreased labor costs and increased productivity.

Reduced labor results because one operator can run the machining center and thereby perform milling, drilling, contouring, and a variety of other operations. In a traditional machining setting, each operation is performed on an individual machine with its own individual

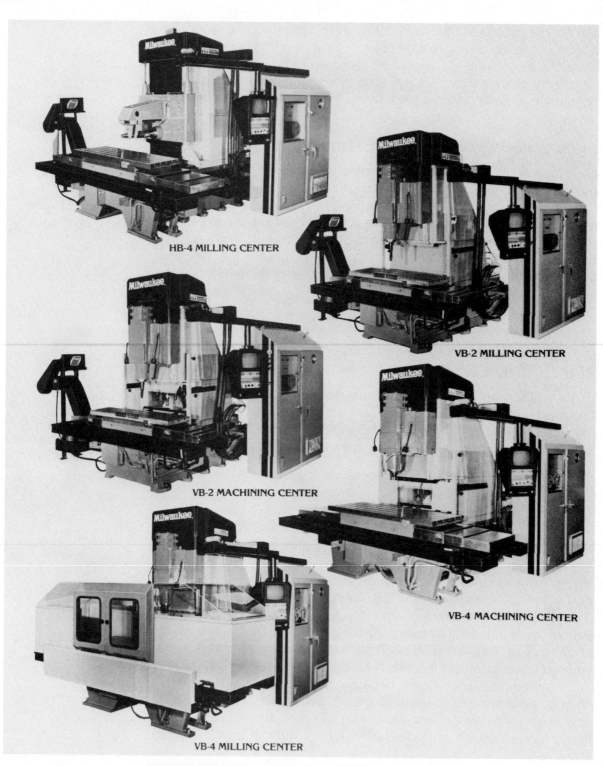

HB-4 MILLING CENTER

VB-2 MILLING CENTER

VB-2 MACHINING CENTER

VB-4 MACHINING CENTER

VB-4 MILLING CENTER

FIGURE 8–9 Machining centers. *Courtesy of Kearney & Trecker.*

operator. Increased productivity results because with a machining center such tasks as setup, tool changing, and loading/unloading take less time than with conventional machine tools. Figures 8–10 and 8–11 are examples of modern machining centers.

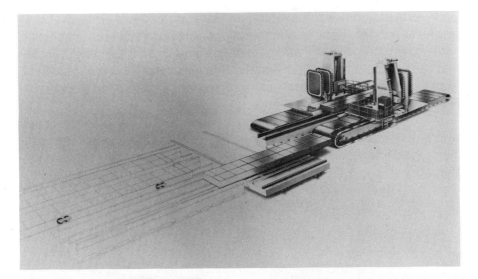

FIGURE 8–10 **Machining center.** *Courtesy of Cincinnati Milacron.* Note: Safety equipment may have been removed or opened to clearly illustrate the product and must be in place prior to operation.

FIGURE 8–11 **Milwaukee-Matic 800 Machining Center.** *Courtesy of Kearney & Trecker.*

Types of Machining Centers

There are two main types of machining centers:

1. vertical spindle
2. horizontal spindle

Both vertical and horizontal spindle machining centers are available in single- or multiple-spindle configurations. Also available are machining centers with both vertical and horizontal spindles or adjustable spindles that can be rotated into either horizontal or vertical positions. Figure 8–12 is an example of a horizontal machining center. Figure 8–13 is an example of a vertical machining center.

Vertical spindle machining centers are used in those cases where downward force is important to the machining process. The vertical spindle machining center is well suited to producing flat or box-shaped workpieces. Horizontal spindle machining centers are used for machining multiple-sided workpieces.

FIGURE 8–12 BPC 320H horizontal machining center. *Courtesy of Bridgeport-Textron.*

FIGURE 8–13 **Vertical machining center.** *Courtesy of Cincinnati Milacron.* Note: Safety
equipment may have been removed or opened to clearly illustrate the product and
must be in place prior to operation.

In addition to vertical and horizontal machining centers, there
are **universal machining centers**. These hybrid centers have the following characteristics:

1. horizontal and vertical spindles with automatic switching capability from one to the other
2. tool changer for the horizontal spindle
3. tool changer for the vertical spindle
4. pallet changer

Figure 8–14 shows where machining centers fit into the manufacturing continuum. Used in conjunction with an FMC or an FMS, machining centers can enhance the productivity and versatility of a manufacturing plant. As progress toward wider use of flexible manufacturing
techniques is made, machining centers will play an increasingly important role.

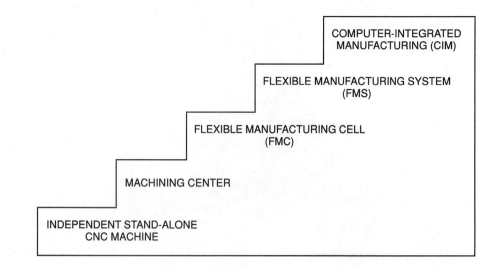

FIGURE 8–14 Manufacturing continuum.

KEY TERMS

Flexible manufacturing
Flexible manufacturing
    system (FMS)
Computer numerical
    control (CNC)
Distributed numerical
    control (DNC)
Automated materials handling
Group technology
System 24
Transfer line

Family of parts
Simultaneous production
Control system
Flexible manufacturing cell
    (FMC)
Artificial intelligence
Machining center
Vertical spindle machining center
Horizontal spindle
        machining center
Universal machining center

Chapter Eight REVIEW

1. What is a flexible manufacturing system (FMS)?
2. What are the key elements that make a manufacturing system an FMS?
3. What four concepts are integrated in an FMS?
4. List and explain at least four tasks humans accomplished in an FMS.
5. Why did the first FMS fail?

6. List the leading American vendors of FMSs.
7. Explain the manufacturing continuum in terms of trade-offs between production rates and flexibility. Where does flexible manufacturing fit in?
8. List and explain at least four advantages of flexible manufacturing.
9. List and explain the four major components of an FMS.
10. List and explain the four functions served by the control system in an FMS.
11. What is a flexible manufacturing cell (FMC)? How does an FMC differ from an FMS?
12. List four advantages of an FMC over an FMS.
13. What is a machining center?
14. What are the two characteristics that distinguish a machining center from a stand-alone CNC machine?
15. Describe the two main types of machining centers.
16. What is a universal machining center?
17. What are the four main characteristics of a universal machining center?

These case studies were provided by the Society of Manufacturing Engineers (SME). They are excerpted from the SME's *Manufacturing Insights*® series of videotapes. These case studies give students actual examples of the way the advanced manufacturing technologies covered in this chapter are being applied in "real-world" manufacturing settings.

As you read each case study, relate the examples cited to the material presented in the text of the chapter. This combination of textbook information and real-world examples will be particularly valuable to your understanding of advanced manufacturing technologies.

AGNEW MACHINE COMPANY

One supplier to the automotive market, with a desire to keep primary manufacturing in the United States by exploiting current technology, is Agnew Machine Company of Highland, Michigan. This systems integrator designed and built a flexible manufacturing cell (FMC) for one of the automotive "Big Three" to machine a family of exhaust manifolds for a V6 engine, including the right-hand manifold, a left-hand manifold, a special right-hand manifold for special engines, and a third right-hand manifold as a service part. Parts are manually loaded into fixtures and clamped with a power wrench. Gross production rates are 20 pairs of standard manifold and 12 service parts every hour.

After parts are manually loaded, a load-assist mechanism is used to make sure the locating pads on the part are firmly in contact with the fixture locaters. At the load station, an identification system using proximity sensors determines part identity. This data is then fed into the computer numerical control (CNC) controller to select the correct machining program for the part. Depending on the part, this may range from a simple milling operation on the joint faces to a complex series of operations including drilling and tapping several holes.

The heart of the CNC machining center is a three-axis traveling column unit, a machine well suited to this production application. It features a recirculating oil system to lubricate and cool the precision cartridge-type spindle.

The Norte machining center is equipped with a 30-tool capacity, stationary toolchanger. This provides ample storage for all necessary tools, plus spares to support long production runs. The toolchanger is totally enclosed, so tools are protected from chips and coolant, as a part of overall chip management to improve uptime of the system. The design of the toolchanger also permits tools to be loaded or unloaded for maintenance with complete safety, even while the machine is operating.

Control is supplied by a Siemens 3M Sinumeric CNC. It has sufficient random access memory to store all four parts programs and such advances a color graphic cathode ray tube (CRT). Programming and control functions such as overrides are easily accomplished by floor-level personnel, without extensive computer training.

The Agnew Dial-A-Flex FMC can be easily incorporated into a more complex system in the future, if desired. Agnew has also pioneered the use of an unusual parts-holding fixture designed to address multisurface flexible machining needs, called the Cuboidal Flex System.

The workpiece is clamped into a hollow cubic fixture, and by manipulating the cube, the workpiece can be approached by any of its six sides for machining. This cell was built for a major automotive manufacturer to machine the rear retainer used on four-wheel-drive transfer cases. The system handles seven versions of this part family. All have common lugs, but different holes of different diameters: speedometer holes, speedometer clips, and various rail holes and mounting holes, a challenge to flexibility.

It all starts at the load-assist station, where automatic torque wrenches tighten the clamps to predetermined values and the cuboidal fixture is released on the conveyor toward the first machining station. The fixture is *not* transported on its locating surface. When it arrives at the station, it is clamped up on the machining cell rather than down, to prevent wear and provide a clean operation for chip management.

When the cell completes the necessary machining, the cube moves on through the system. In this case, after the workpiece has been through the second cell, the fixture passes over a rollover station, which rotates it 90 deg to position it for the next station.

Obviously, with seven different parts to be machined, a great deal of control sophistication is built into the system. Much of this is handled by the Giddings & Lewis PC409 programmable controllers, which serve in both a supervisory function and as cell controllers, and the Norte-equipped cells using Seimens 3M Sinumeric CNC control.

(continued)

Worldwide competitive pressures, coupled with the industrial restructuring, have made FMCs necessary. However, it is recent technological developments that have made them possible.

YAMAZAKI MAZAK

Both the competitive and international aspect of the flexible manufacturing cell (FMC) is very evident at Mazak Sales and Service in Florence, Kentucky. The Florence plant is part of Yamazaki Mazak of Oguchi, Japan, a world leader in flexible manufacturing equipment with a network of manufacturing plants, sales and service locations, and technical centers across the globe. The Florence plant has the capacity to put out well over 100 machining centers in a month's time, allowing the exportation of approximately 30% of its current production.

Mazak's commitment to quality is immediately evident on viewing its rigidly maintained clean-room environments for such critical operations as spindle and bearing assembly. Evidence of Mazak's commitment to the U.S. market is its plan to expand production over the next few years, including a broader range of horizontal machining centers and larger vertical machining centers.

At Mazak they "practice what they preach" by utilizing their own flexible machining equipment to produce contemporary FMCs for a wide range of applications and they do it with innovative technology in a virtually unmanned, "lights-out" environment. Their flexible machining centers have been set up to perform multiple machining operations on many types of Mazak castings. These centers include a floor space-saving multiple-pallet changer system that can be expanded from 15 to 60 with the addition of more pallet stockers and an expandable "tool hive" for tool storage that includes the automatic transfer of tools from storage pockets to ATC position. The basic 160-tool capacity can be increased up to 480.

Control of machining center operations is handled by the central processing unit (CPU). On the CPU, schedule data can be input and modified, and the location and status of all pallets is displayed in real time. The required program number, time scheduled to complete an operation, and accumulated total of uncompleted machining can be displayed on a terminal.

The selection of tools, monitoring of cutting conditions, and generation of optimum tool paths is handled by the individual Mazatrol horizontal machining center controllers. Current machining operations can be displayed on the operator panel's high-resolution color monitor.

Mazak has found that the FMC concept is rapidly gaining popularity with its customers. Brian J. Papke, Vice President/General Manager says:

> There is a recognition on the part of customers that where they used to use dedicated equipment for machining their parts, it is no longer practical to do that because they have to be able to respond to changes in engineering and changes in part mix so that they must have more versatile equipment than they used previously. Also, as they look at that process, there's a tremendous emphasis in companies today on reducing their inventory. Certainly, one good method is to reduce the times that you take a machine off a process.
>
> Obviously, when we talk about cells for many customers, the emphasis is still reducing the overall time necessary to make the part by investing in versatile equipment that is state of the art. In making that part, they can overall reduce the amount of manpower that is required to make that part, because they can run in an unmanned environment over several machines. Certainly in our own factories, that is true.

Thus, the Flexible Manufacturing Cell concept has been very successful—both as a vendor and a user—for Mazak.

Major Topics Covered

- CIM Defined
- Historical Development of CIM
- Problems Associated with CIM
- The "CIM Wheel"
- Benefits of CIM
- CIM-Related Standards
- Just-In-Time and CIM
- MRP and CIM
- Artificial Intelligence and CIM
- *Case Study: Mack Truck*
- *Case Study: LTV Aerospace and Defense*

Chapter Nine

Automated Assembly

Automated Guided Vehicles

CAD/CAM

Numerical Control

Industrial Robots

Lasers in Manufacturing

Programmable Logic Controllers

Flexible Manufacturing

Computer Integrated Manufacturing

Other Related Technologies

CIM DEFINED

Computer-integrated manufacturing or (CIM) is the term used to describe the most modern approach to manufacturing. Although CIM encompasses many of the other advanced manufacturing technologies such as computer numerical control (CNC), computer-aided design/computer-aided manufacturing (CAD/CAM), robotics, and just-in-time delivery (JIT), it is more than a new technology or a new concept. Computer-integrated manufacturing is actually an entirely new approach to manufacturing, a new way of doing business.

To understand CIM, it is necessary to begin with a comparison of modern and traditional manufacturing. Modern manufacturing encompasses all of the activities and processes necessary to convert raw materials into finished products, deliver them to the market, and support them in the field. These activities include the following (Figure 9–1):

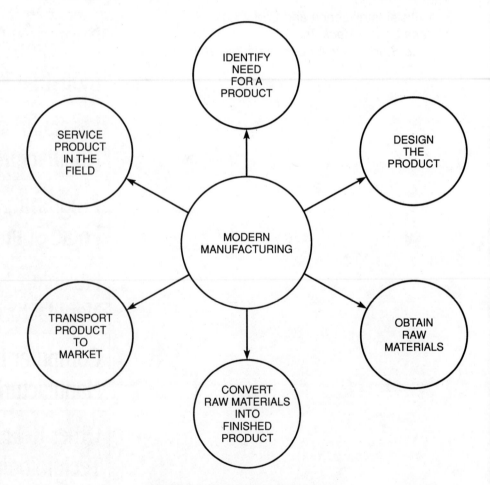

FIGURE 9–1 Modern manufacturing steps.

1. identifying a need for a product
2. designing a product to meet the needs
3. obtaining the raw materials needed to produce the product
4. applying appropriate processes to transform the raw materials into finished products.
5. transporting product to the market
6. maintaining the product to ensure a proper performance in the field

This broad, modern view of manufacturing can be compared with the more limited traditional view that focuses almost entirely on the conversion processes. The old approach separates such critical pre-conversion elements as market analysis research, development, and design for manufacturing, as well as such after-conversion elements as product delivery and product maintenance. In other words, in the old approach to manufacturing, only those processes that take place on the shop floor are considered manufacturing. This traditional approach of separating the overall concept into numerous stand-alone specialized elements was not fundamentally changed with the advent of automation. While the separate elements themselves became automated (i.e., computer-aided drafting and design (CADD) in design and CNC in machining), they remained separate. Automation alone did not result in the integration of these **islands of automation**.

With CIM, not only are the various elements automated, but the islands of automation are all linked together or integrated. **Integration** means that a system can provide complete and instantaneous sharing of information. In modern manufacturing, integration is accomplished by computers. With this background, CIM can now be defined as

THE TOTAL INTEGRATION OF ALL MANUFACTURING ELE-MENTS THROUGH THE USE OF COMPUTERS.

Figure 9–2 is an illustration of a CIM system, which shows how the various machines and processes used in the conversion process are integrated. However, such an illustration cannot show that research, development, design, marketing, sales, shipping, receiving, management, and production personnel all have instant access to all information generated in this system. This is what makes it a CIM system.

HISTORICAL DEVELOPMENT OF CIM

The term "computer-integrated manufacturing" was developed in 1974 by Joseph Harrington as the title of a book he wrote about tying islands of automation together through the use of computers. It has

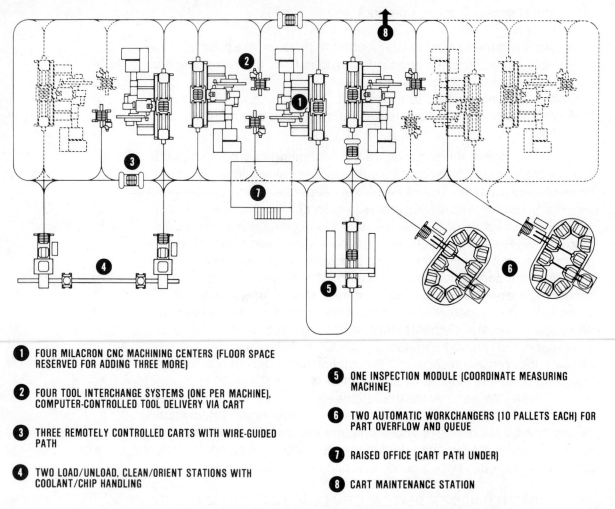

1. **FOUR MILACRON CNC MACHINING CENTERS (FLOOR SPACE RESERVED FOR ADDING THREE MORE)**

2. **FOUR TOOL INTERCHANGE SYSTEMS (ONE PER MACHINE), COMPUTER-CONTROLLED TOOL DELIVERY VIA CART**

3. **THREE REMOTELY CONTROLLED CARTS WITH WIRE-GUIDED PATH**

4. **TWO LOAD/UNLOAD, CLEAN/ORIENT STATIONS WITH COOLANT/CHIP HANDLING**

5. **ONE INSPECTION MODULE (COORDINATE MEASURING MACHINE)**

6. **TWO AUTOMATIC WORKCHANGERS (10 PALLETS EACH) FOR PART OVERFLOW AND QUEUE**

7. **RAISED OFFICE (CART PATH UNDER)**

8. **CART MAINTENANCE STATION**

FIGURE 9–2 Modern CIM system. *Courtesy of Cincinnati Milacron.*

taken many years for CIM to develop as a concept, but integrated manufacturing is not really new. In fact, integration is where manufacturing actually began. Manufacturing has evolved through four distinct stages (Figure 9–3):

1. manual manufacturing
2. mechanization/specialization
3. automation
4. integration

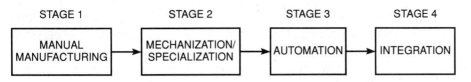

STAGE 1	STAGE 2	STAGE 3	STAGE 4
MANUAL MANUFACTURING	MECHANIZATION/ SPECIALIZATION	AUTOMATION	INTEGRATION

FIGURE 9–3 Historical development of integrated manufacturing.

Manual Manufacturing

Manual manufacturing using simple hand tools was actually integrated manufacturing. All information needed to design, produce, and deliver a product was readily available because it resided in the mind of the one person who performed all of the necessary tasks. The tool of integration in the earliest years of manufacturing was the human mind of the craftsman who designed, produced, and delivered the product. An example of integrated manual manufacturing is the village blacksmith producing a special tool for a local farmer. The blacksmith would have in his mind all of the information needed to design, produce, and deliver the farmer's tool. In this example, all elements of manufacturing are integrated.

Mechanization/Specialization

With the advent of the industrial revolution, manufacturing processes became both specialized and mechanized. Instead of one person designing, producing, and delivering a product, workers and/or machines performed specialized tasks within each of these broad areas. Communication among these separate entities was achieved using drawings, specifications, job orders, process plans, and a variety of other communication aids. To ensure that the finished product matched the planned product, the concept of quality control was introduced.

The positive side of the **mechanization/specialization** stage was that it permitted mass production, interchangeability of parts, entire levels of accuracy, and uniformity. The disadvantage is that the lack of integration led to a great deal of waste.

Automation

Automation improved the performance and enhanced the capabilities of both people and machines within specialized manufacturing components. For example, CADD enhanced the capability of designers and drafters; CNC enhanced the capabilities of machinists; and **computer-assisted process planning (CAPP)** enhanced the capabilities of industrial planners. But the improvements brought on by automation were isolated within individual components or islands. Because of this, automation did not always live up to its potential.

To understand the limitations of automation with regard to overall productivity improvement, consider the following analogy. Suppose that various subsystems of an automobile (i.e., engine, steering, brakes) were automated to make the driver's job easier. Automatic acceleration, deceleration, steering, and braking would certainly be more efficient than the manual versions. However, consider what would happen if these various automated subsystems were not tied together in a way that allowed them to communicate and share accurate, up-to-date information instantly and continually. One system might be attempting to accelerate the automobile while another system was attempting to apply the brakes. The same limitations apply in an automated manufacturing setting. These limitations are what led to the current stage in the development of manufacturing: integration.

Integration

With the advent of the computer age, manufacturing has developed full circle. It began as a totally integrated concept and, with CIM, has once again become one. However, there are major differences in the manufacturing integration of today and that of the manual era of the past. First, the instrument of integration in the manual era was the human mind. The instrument of integration in modern manufacturing is the computer. Second, processes in the modern manufacturing setting are still specialized and automated.

Another way to view the historical development of CIM is by examining the ways in which some of the individual components of CIM have developed over the years. Such components as design, planning, and production have evolved both as processes and in the tools and equipment used to accomplish the processes.

Design has evolved from a manual process using such tools as slide rules, triangles, pencils, scales, and erasers into an automated process known as computer-aided design (CAD). Process planning has evolved from a manual process using planning tables, diagrams, and charts into an automated process known as computer-aided process planning (CAPP). Production has evolved from a manual process involving manually controlled machines into an automated process known as computer-aided manufacturing (CAM).

These individual components of manufacturing evolved over the years into separate islands of automation. However, communication among these islands was still handled manually. This limited the level of improvement in productivity that could be accomplished in the overall manufacturing process. When these islands and other automated components of manufacturing are linked together through computer networks, these limitations can be overcome. Computer-integrated manufacturing has enormous potential for improving productivity in manufacturing, but it is not without problems.

PROBLEMS ASSOCIATED WITH CIM

As with any new philosophy that requires major changes to the status quo, CIM is not without problems. The problems associated with CIM fall into three major categories (Figure 9–4):

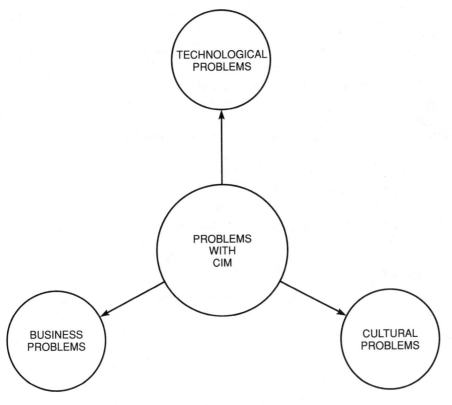

FIGURE 9–4 Problems with CIM.

1. technical problems
2. cultural problems
3. business-related problems

These types of problems have hindered the development of CIM over the years and will have to be overcome for CIM to achieve widespread implementation.

Technical Problems of CIM

As each island of automation began to evolve, specialized hardware and software for that island were developed by a variety of producers. This led to the same type of problem that has been experienced in the automotive industry. One problem in maintaining and repairing automobiles has always been the incompatibility of spare parts among various makes and models. **Incompatibility** summarizes in a word the principal technical problem inhibiting the development of CIM. Consider the following example. Supplier A produces hardware and software for automating the design process. Supplier B produces hardware and software for automating such manufacturing processes as machining, assembly, packaging, and materials handling. Supplier C produces hardware and software for automating processes associated with market research. This means a manufacturing firm may have three automated components, but on systems produced by three different suppliers. Consequently, the three systems are not compatible. They are not able to communicate among themselves. Therefore, there can be no integration of the design, production, and market research processes.

An effort known as manufacturing automation protocol (MAP) is beginning to solve the incompatibility of hardware and software produced by different suppliers. As MAP continues to evolve, the incompatibility problem will eventually be solved and full integration will be possible among all elements of a manufacturing plant.

Cultural Problems of CIM

Computer-integrated manufacturing is not just new manufacturing technology; it is a whole new approach to manufacturing, a new way of doing business. As a result, it involves significant changes for people who were educated and are experienced in the old ways. As a result, many people reject the new approach represented by CIM, for a variety of reasons. Some simply fear the change that it will bring in their working lives. Others feel it will altogether eliminate their positions, leaving them functionally obsolete. In any case, the cultural problems associated with CIM will be more difficult to solve than the technical problems.

Business-Related Problems of CIM

Closely tied to the cultural problems are the business problems associated with CIM. Prominent among these is the accounting problem. Traditional accounting practices do not work with CIM. There is no way to justify CIM based on traditional accounting practices. Traditional accounting practices base cost-effectiveness studies on direct labor savings whenever a new approach or new technology is proposed. However, the savings that result from CIM are more closely tied to indirect and intangible factors, which are more difficult to quantify. Consequently, it can be difficult to convince traditional business people, who are used to relying on traditional accounting practices, to see that CIM is an approach worth the investment.

THE CIM WHEEL

The Computer and Automated Systems Association (CASA) of the Society of Manufacturing Engineers (SME) developed the **CIM wheel** (Figure 9–5) as a way to comprehensively, but concisely illustrate the concept of CIM. The CASA/SME developed the CIM wheel to include five distinct components (Figure 9–6):

1. general business management
2. product and process definition
3. manufacturing planning, and control
4. factory automation
5. information resource management

General Business Management

There are four principal elements of the general business management component of the CASA/SME CIM wheel. These four elements are shaded on the CIM wheel in Figure 9–7. They encompass all of those activities associated with doing any kind of business. They link the rest of the components of the CIM wheel to the outside world. Many of the processes within these four elements are automated. In manufacturing firms that have moved forward with automation, these four elements of the general business management component typically become individual islands of automation.

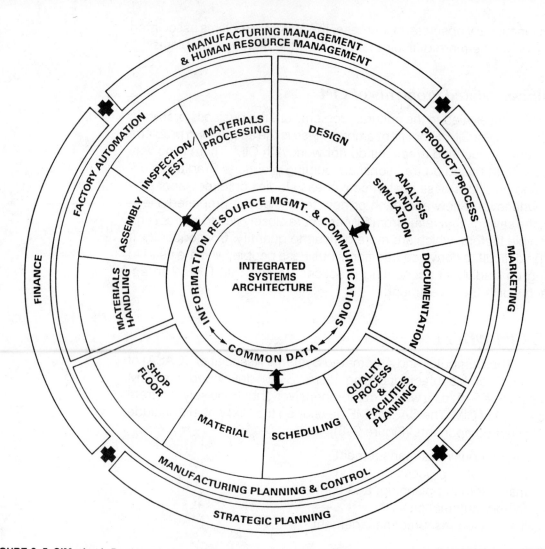

FIGURE 9–5 CIM wheel. *Reprinted courtesy of the Computer and Automated Systems Association (CASA) of the Society of Manufacturing Engineers.*

Within the finance element, it is not uncommon to have an automated payroll system, an automated accounts receivable system, and an automated accounts payable system. Such automation also exists in the other three components. However, in the typical automated manufacturing firm, these islands of automation within the general business management component are not networked for integration even within the component, much less with the other elements of the CIM wheel. For CIM to exist, every component within the CIM wheel must be networked

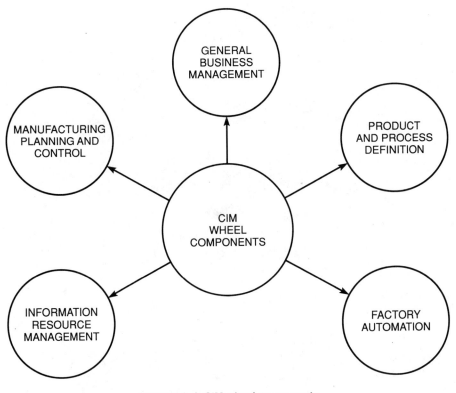

FIGURE 9–6 CIM wheel components.

and every individual element within components must be networked for instantaneous exchanges and updates of data.

Product and Process Definition

The **product and process definition** component of the CIM wheel contains three elements. These three elements are shaded in the CIM wheel in Figure 9–8:

1. design
2. analysis and simulation
3. documentation

This is the component in which a product is engineered, designed, tested through simulation, documented through drawings, specifications, and other documentation tools such as parts lists and bills of material. Islands of automation for this component of the CIM wheel have been emerging since the late 1960s.

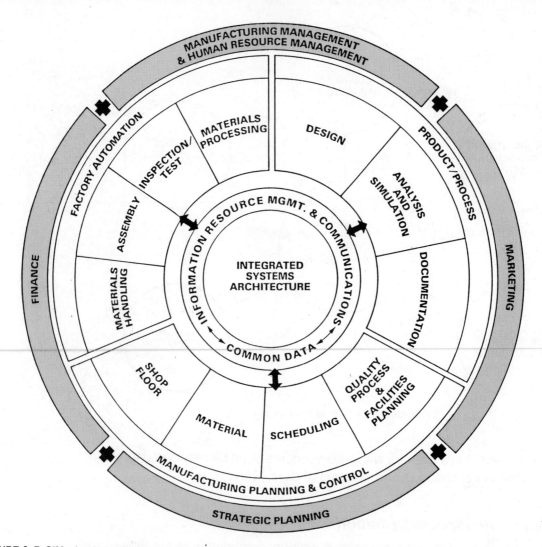

FIGURE 9–7 CIM wheel showing elements of the general business management component. *Reprinted courtesy of the Computer and Automated Systems Association (CASA) of the Society of Manufacturing Engineers.*

These islands include CADD systems, modeling and simulation software including solids modeling, surface modeling, and finite-element analysis. Also within this component are such islands of automation as CAPP. Even in highly automated manufacturing plants in which all of the product and process definition systems are automated, it is rare to find effective networking and integration of the processes within this individual component, much less among the various other components of the CIM wheel. It is not uncommon, within this component of the wheel, to

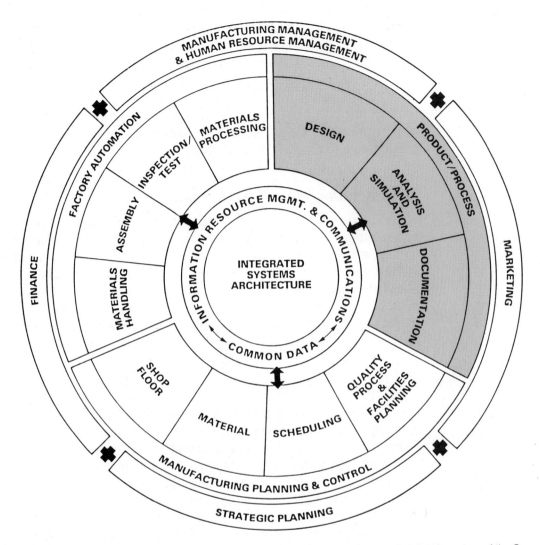

FIGURE 9–8 CIM wheel showing elements of the product and process component. *Reprinted courtesy of the Computer and Automated Systems Association (CASA) of the Society of Manufacturing Engineers.*

find incompatible hardware and software being used even in individual elements of a component such as design.

An example of this would be a company that automated its design processes early by purchasing hardware and software from supplier A. Later as technology continued to evolve, supplier B produced a better system and the company purchased it. However, due to financial limitations, the company was not able to purchase as many stations of the new system as it needed. As a result, some engineers and designers continue to work on the old automated system while others work at new stations. Because of differences in the hardware and software produced by the two suppliers, the old and new systems are incompatible. As a result, not only can this company not network its design functions with other components on the CIM wheel, it cannot even network within the product and process definition component. This type of incompatibility is more often the rule rather than the exception. It represents the principal obstacle to the full development of CIM.

Manufacturing Planning and Control

The **manufacturing planning and control** component includes such elements as facilities planning, scheduling, material planning and control, and shop floor planning and control. These elements are shaded in the CIM wheel in Figure 9–9. Hardware and software are available to automate each of the individual elements within this component. However, as with the previous group, there is rarely integration of the elements within this component, much less outside of it. The chief problem here is also incompatibility.

Factory Automation

The **factory automation** component contains those elements normally associated with producing the product: materials handling, assembly, inspection and testing, and materials processing. This component is shaded in Figure 9–10. Much of the research and development in the area of automated manufacturing has focused on this group. Such automated manufacturing concepts as CNC, distributed numerical control (DNC), industrial robots, and automated materials handling systems such as automated guided vehicles (AGVs) have been available for over 20 years. During this time, they have continually improved in performance. However, very little progress has been made in successfully networking the elements within this group with those outside of it. Some progress is being made through the concept of CAD/CAM in which the product and process definition islands of automation are networked with the factory automation elements. However, incompatibility remains the key inhibitor to full integration.

Information Resource Management

The **information resource management** component of the CIM wheel is located in the center of Figure 9–11. This is an appropriate

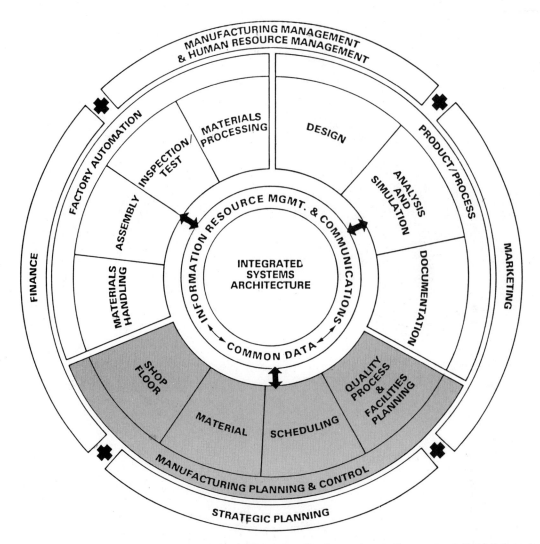

FIGURE 9-9 CIM wheel showing elements of the manufacturing planning and control component. *Reprinted courtesy of the Computer and Automated Systems Association (CASA) of the Society of Manufacturing Engineers.*

position for this component because it represents the nucleus of CIM. Information, updated continually and shared instantaneously, is what CIM is all about. To integrate the various elements within the various components of the CIM wheel, all of the information generated by the various components must be effectively managed. One of the major goals of this component is to overcome the barriers that prevent the complete sharing of information between and among components in the CIM wheel.

There are two basic elements within this component: the information being managed and the hardware and software used to manage

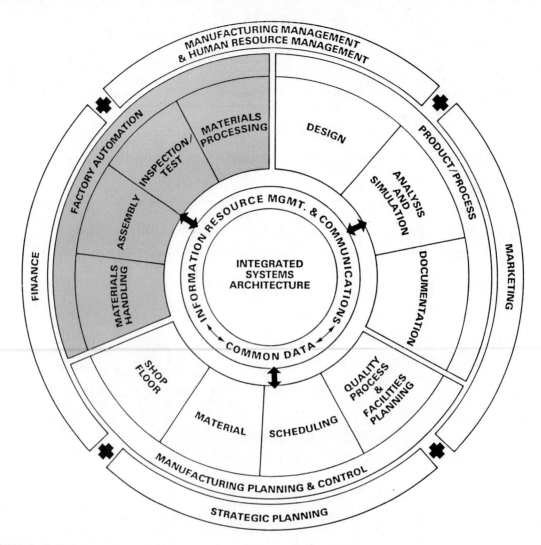

FIGURE 9–10 CIM wheel showing elements of the factory automation component. *Reprinted courtesy of the Computer and Automated Systems Association (CASA) of the Society of Manufacturing Engineers.*

that information. The technology used to manage information within this component can be divided into four categories by function (Figure 9–12):

1. communications technology
2. network transaction technology
3. data management technology
4. user technology

Each of these elements represents a different layer of computer technology. Achieving full integration of all elements and all components

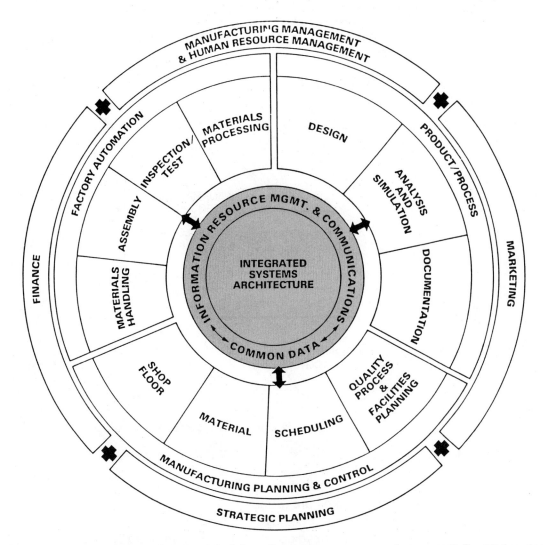

FIGURE 9–11 CIM wheel showing elements of the information resource management component. *Reprinted courtesy of the Computer and Automated Systems Association (CASA) of the Society of Manufacturing Engineers.*

of the CIM wheel involves successfully horizontal and vertical networking at all four of these levels.

BENEFITS OF CIM

In spite of the obstacles, progress is being made toward the eventual full realization of CIM in manufacturing. When this is accomplished, fully integrated manufacturing firms will realize a number of benefits from CIM:

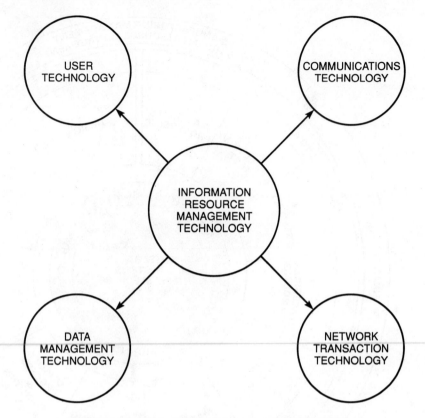

FIGURE 9–12 Information resource management technology.

1. product quality increases
2. lead times are reduced
3. direct labor costs are reduced
4. product development times are reduced
5. inventories are reduced
6. overall productivity increases
7. design quality increases

CIM-RELATED STANDARDS

The point has been made repeatedly in this chapter that incompatibility is the principal inhibitor of the full development of CIM. Standards are currently being developed to help overcome the problem of incompatibility. Three in particular are beginning to have a positive impact (Figure 9–13):

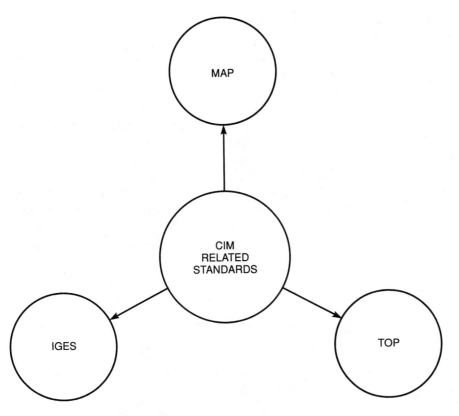

FIGURE 9–13 CIM-related standards.

1. manufacturing automation protocol (MAP)
2. technical and office protocol (TOP)
3. initial graphics exchange specification (IGES)

Manufacturing Automation Protocol

Manufacturing automation protocol (MAP) is a communications standard being developed to promote compatibility among automated manufacturing systems produced by different vendors. It was originally developed by the General Motors Corporation to help them increase productivity so that they could be competitive with foreign automobile manufacturing firms. There are now internationally MAP user groups that have input into the ongoing development and improvement of MAP.

The first version of MAP was published in 1982. Since that time it has been continually updated, improved, and revised. Manufacturing automation protocol is now being used by companies to integrate automated manufacturing systems produced by different vendors. The actual physical link between systems with MAP is a coaxial cable containing a token bus, broad band communications network. Manufacturing automation protocol is based on the open systems interconnection (OSI) seven-layer model (Figure 9–14). This model was adopted by General Motors Corporation because it already had achieved a high degree of acceptance worldwide. Having a broad base of support and acceptance is critical to the success of a proposed standard.

The overall goal of MAP is the total integration of islands of automation in manufacturing, regardless of the producer of the hardware and software used in the system. With MAP fully developed and in place, a user will have access to any computer within a manufacturing facility from any other computer within that facility, regardless of the make, model, or vendor of the computer.

MAP SPECIFICATIONS BY LAYER		
Layer Number	Layer Description	Layer Function
7	Application	Provides all services directly understandable to application programs.
6	Presentation	Converts data to and from agreed to standardized formats.
5	Session	Synchronizes and manages data.
4	Transport	Provides transparent data transfer from end node to end note.
3	Network	Provides pocket routing for transferring data between nodes on different networks.
2	Data Link	Detects errors for messages transferred between nodes on the same networks.
1	Physical	Encodes bits and transfers them between adjacent nodes.

FIGURE 9–14 Manufacturing automation protocol.

Technical and Office Protocol

Technical and office protocol (TOP) is a standard that was developed to promote integration within an office environment. While MAP is used to promote integration among the manufacturing processes contained in components 2 through 5 of the CIM wheel (Figure 9–15), TOP was developed to promote integration in the general business management (Figure 9–16). Boeing has played the lead role in the development of TOP. The purpose of TOP is to allow islands of automation within the first component of the CIM wheel to interface and communicate not just among themselves but also between and among those islands contained in the other four components of the wheel. Technical and office protocol is also based on a seven-layer model, similar to the model illustrated in Figure 9–14.

Technical and office protocol is being developed in three phases. The first phase includes standards for file transfer and, on a limited basis, file management. Phase two includes standards covering file

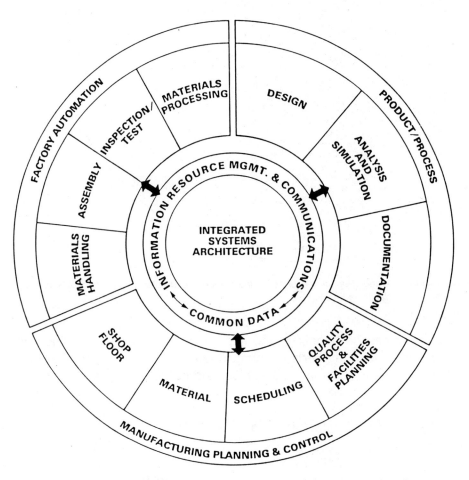

FIGURE 9–15 CIM wheel. *Reprinted courtesy of the Computer and Automated Systems Association (CASA) of the Society of Manufacturing Engineers.*

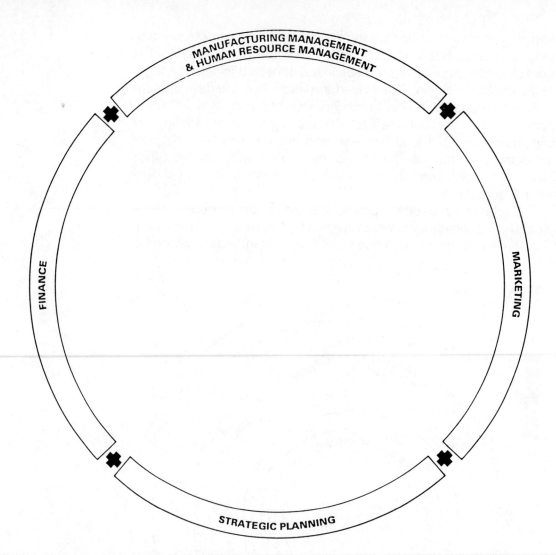

FIGURE 9–16 **First component of the CIM wheel.** *Reprinted courtesy of the Computer and Automated Systems Association (CASA) of the Society of Manufacturing Engineers.*

access, message handling, and improved file management. The final phase includes standards covering document revision, document exchange, directory services, graphics, and database management.

Initial Graphics Exchange Specification

The **initial graphics exchange specification (IGES)** is much more limited than MAP or TOP. It was developed to promote communication between CADD systems manufactured by different vendors and is limited to the product and process definition component of the CIM wheel. The central technology of an IGES specification is a translator that is arranged between two CADD systems that are communicating. System A transfers the data that is to be moved to system B into this translator, where it is converted into the neutral IGES format. It then is converted from that format into the format understood by the CADD system B (Figure 9-17).

A key weakness of IGES is its inability to deal with three-dimensional solid models, electronic design, and nongraphic data management. The National Bureau of Standards is currently developing a new standard called the product data exchange standard (PDES), that will be able to handle solids modeling data, nongraphic data management, and electronic design.

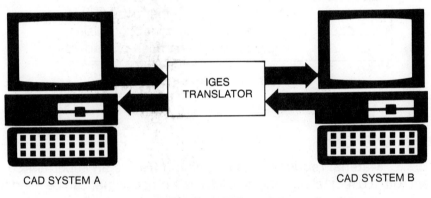

CAD SYSTEM A   CAD SYSTEM B

FIGURE 9-17 IGES translator.

JUST-IN-TIME AND CIM

Just-in-time (JIT) is a concept closely associated with CIM. It is typically thought of as an inventory program. However, this is not a complete definition. Like CIM, JIT is actually more than just a system; it is a new operating philosophy. Although JIT is generally thought of as a new concept, it was actually born in the 1920s at Ford Motor Company. The basic idea of JIT is the same now as it was then: reduce production cycle time while eliminating waste.

At the heart of the JIT philosophy is the concept of "value added." Anything that does not add value to a given product is considered waste. For example, large amounts of raw materials sitting in a warehouse does not add value to the finished product. With a JIT system, there is no large inventory of materials in warehouses. The materials needed are delivered from the supplier to the manufacturer where and when they are needed "just in time." This is why so many people think of JIT as a new inventory control system.

The overall goal of a JIT system is to identify and eliminate any aspect of any process involved in producing a product that does not add value to the product. Figure 9–18 is a summary of the essentials of JIT.

JUST-IN-TIME CHECKLIST	
✔	Stable final assembly production
✔	Limited batch production
✔	Shorter supply lead times
✔	Shorter production lead times
✔	Better quality
✔	Easily integrated with MRP
✔	Decreases the number of suppliers
✔	Greater at source control of quality
✔	Improves documentation procedures

FIGURE 9–18 Overview of JIT.

MRP AND CIM

Material requirements planning (MRP) is an important concept with a direct relationship to CIM. It is a process that can be used to calculate the amount of raw materials that must be obtained in order to manufacture a specified lot of a certain product. Material requirements planning involves using the bill of material, production schedule, and inventory record to produce a comprehensive, detailed schedule of the raw materials and components needed for a job.

As manufacturing technology has evolved from automation to integration, MRP has also evolved. The acronym **MRP** now means **manufacturing resource planning**. This broader concept goes beyond determining materials requirements to also encompass financial tracking and accounting. The modern version of MRP is particularly well suited to the integrated approach represented by CIM. In this approach, MRP can be an effective inventory planning and control tool.

Key concepts relating to MRP include (a) independent and dependent demand, (b) lead times, and (c) common-use items. Independent demand resources are those that are not tied to any other resource. They stand alone. Dependent demand resources are tied directly to other resources. Receiving a dependent resource without the other resources it requires does no good. In manufacturing, resources are more likely to be dependent than independent. Raw materials, in-progress parts, components, and subassemblies are all part of the overall manufacturing inventory.

Lead times are of two types: (a) ordering lead time and (b) manufacturing lead time. Ordering lead time is the total amount of time between initiating a purchase order and receiving the order. Manufacturing lead time is the total amount of time required to perform all the steps necessary to produce a given part.

Lead times are important because they are used in developing the schedules for ordering materials and producing products. They are also were MRP is most likely to break down. Resource planners depend on the lead times provided to them by other personnel. If these times are padded or inaccurate, the MRP results will be equally inaccurate.

Common-use items are items used in producing more than just one product. For example, the same type of aluminum sheet might be used in producing several different parts. The integrated approach of MRP allows planners to identify common-use items for a number of different products and use this information to save money by ordering in quantity.

Manufacturing resource planning results in a variety of products of value to manufacturing managers in addition to the master schedule:

1. release notices that notify the purchasing department to place orders
2. revised schedules showing updated due dates
3. cancellation notices that notify appropriate personnel of cancellations that result from changes to the master schedule
4. inventory status reports

Manufacturing resource planning is the most appropriate planning approach for a CIM setting. When completely implemented it can result in a number of benefits:

1. inventory reduction
2. quicker response to demand changes
3. reductions in setup costs
4. more efficient machine utilization
5. quicker response to revisions to the master schedule

ARTIFICIAL INTELLIGENCE AND CIM

A technology that will help in the full development of CIM is artificial intelligence. A most difficult problem to overcome in the factory automation component of CIM is the inability of production systems to mimic such basic human capabilities as adjusting appropriately to differences in the size, shape, and/or orientation of objects. Artificial intelligence will help solve this dilemma.

For example, in a traditional factory setting, an assembly worker whose job is to retrieve small parts from a feeder bin and insert them into the appropriate holes in a plate has many human attributes to assist him. These include sight, hand-eye orientation, reasoning abilities, logic, judgment, and experience. If the assembly worker picks up a part that is not properly oriented or if on his first attempt to insert the part it does not properly seat, these human capabilities allow him to adjust appropriately. Even the most sophisticated automated assembly systems cannot completely mimic these capabilities. Artificial intelligence is an attempt to increase the number of human characteristics computers and computer-controlled systems can mimic.

Artificial Intelligence Defined

Artificial intelligence is the ability of a computer to imitate human intelligence and, thereby, make intelligent decisions. Computer-controlled systems that apply artificial intelligence to everyday settings are called expert systems.

In any discussion of artificial intelligence, there are several key words and phrases that are frequently used. Students of automated assembly should be familiar with at least the following artificial intelligence related terms:

Algorithm: A special computer program that will solve selected problems within a given time frame.

Early Vision: Computer calculations that allow systems to see by providing low-level data such as spatial and geometrical information.

Higher Level Vision: Computer calculations that allow systems to accomplish higher level tasks such as smart movement within an environment, object recognition, and reasoning about objects.

Knowledge Engineering: A process through which knowledge is collected from experts in a given field and converted into a computable format.

Neurocomputing: An approach to performing calculations on a computer that is based on how the human nervous system operates.

Humans are able to make logical, reasoned adjustments in a work setting because they are able to quickly collect information, access it against the sum total of their human experience, and evaluate known relationships among various items of information. The science of artificial intelligence attempts to imitate this process with computers.

Humans attempt to create an experience in computers by feeding them all known information about a given subject. This information is then used by the computer in making decisions. This is why the concept is called artificial intelligence. The computer does not really think. Rather, it simply searches its memory for the appropriate information. If the information is there, the computer uses it in making logical decisions. The key is in feeding the computer enough relevant information.

Artificial Intelligence in CIM

Artificial intelligence has applications in a number of manufacturing settings, one of which is CIM. In such a setting, artificial intelligence can be used to allow automated systems such as robots to duplicate such human capabilities as vision and language processing. It can help improve the manufacturing skills of robots and other machines. Finally, artificial intelligence can improve the ability of information management systems used in CIM settings.

KEY TERMS

Computer-integrated
          manufacturing (CIM)
Islands of automation
Integration
Manual manufacturing
Mechanization/specialization
Automation
Computer-aided process
          planning (CAPP)
Incompatibility
CIM wheel
Product and process definition
Manufacturing planning
          and control

Factory automation
Information resource
          management
Manufacturing automation
          protocol (MAP)
Technical and office
          protocol (TOP)
Initial graphic exchange
          specification (IGES)
Just-in-time (JIT) delivery
Manufacturing resource
          planning (MRP)
Artificial intelligence

Chapter Nine REVIEW

1. Define the term computer integrated manufacturing.
2. Explain the term island of automation.
3. Explain the term integration.
4. Explain the origin of the term computer-integrated manufacturing (CIM).
5. List and explain the four stages in the evolution of manufacturing.
6. Explain the technological problems associated with CIM.
7. Explain the cultural problems associated with CIM.
8. Explain the business-related problems associated with CIM.
9. Explain the general business management component of the SME CIM wheel.
10. Explain the product and process definition component of the SME CIM wheel.
11. Explain the manufacturing, planning, and control component of the SME CIM wheel.
12. Explain the factory automation component of the SME CIM wheel.
13. Explain the information resource management component of the SME CIM wheel.
14. List at least five benefits of CIM.
15. Explain the following CIM-related standards:
    MAP
    TOP
    IGES
16. Explain the concept of just-in-time.

These case studies were provided by the Society of Manufacturing Engineers (SME). They are excerpted from the SME's *Manufacturing Insights*® series of videotapes. These case studies give students actual examples of the way the advanced manufacturing technologies covered in this chapter are being applied in "real-world" manufacturing settings.

As you read each case study, relate the examples cited to the material presented in the text of the chapter. This combination of textbook information and real-world examples will be particularly valuable to your understanding of advanced manufacturing technologies.

MACK TRUCK

A new family of aluminum transmission housings is produced economically using a classical flexible manufacturing system (FMS) with some interesting features.

Several years ago, Mack Truck in Hagerstown, Maryland, was faced with a tough decision—produce a new family of transmission housings using the conventional transfer line approach or switch to FMS. The flexible manufacturing system got the nod and today a full-powered flexible machining system is up and running in Hagerstown.

Mack's new FMS is used to machine a variety of similar transmission housings and compound boxes required for production of a new transmission. All workpieces are Type 322 aluminum. The working area required is a 30-in. cube. The system currently has capacity for machining about 34,000 transmission housings and about 29,000 compound cases yearly.

In the past, such workpieces would have been produced using conventional transfer line equipment. But for this project, Mack elected to use FMS technology for improved flexibility. Part-print changes are common in transmission manufacturing; with FMS, changes can be made quickly and inexpensively. With a dedicated transfer line, part-print changes are often costly and time consuming.

Another consideration made FMS even more attractive. As sales of the new transmission increased, production of existing transmissions, produced on transfer lines would decline. At some point in the future, the new line would be required to run the new transmission as well

(continued)

as service parts for the older transmissions. Even though a transfer line system would have been more attractive for short-term return on investment, Mack decided to go with FMS as an investment in future flexibility.

Kearney & Trecker built Mack's FMS. It consists of four Milwaukee-matic 800 computer numerical control (CNC) machining centers, provisions for two additional Milwaukee-matics to be installed in the future, two Milwaukee number 24 headchangers, and a special Milwaukee precision duplex boring machine. Three load and unload areas are used to feed workpieces to the system. The materials handling network is a tow-chain type equipped with 20 carts. Chain speed is 50 ft per minute.

The Milwaukee-matics used in the system can handle workpieces within a 39-in. cube, weighing up to 5000 lbs. Equipped with Kearney & Trecker Gemini D-17 CNC units, the 20-horsepower machines have storage capacity for up to 68 tools. Features include spindle probe, adaptive control, and broken tool sensing.

The Milwaukee headchangers perform drilling, tapping, and boring. A ten-head carousel carries 24- x 24-in. multiple-spindle heads. The heads are driven by a 40-horsepower work spindle. Transmission cases require two passes through the system. Workpieces are manually loaded in a vertical fixture. When secured in the fixture, the parts are automatically routed to any one of the Milwaukee-matics. A three-minute qualifying operation is performed in which the top cover face is milled and two dowel holes are drilled. After processing, the part is automatically sent back to a load and unload station.

At this point, the part is removed, manually moved to another load station, then secured in a horizontal fixture for further processing. The milled face and dowel holes produced in the first operation are used for locating. The part is routed automatically to any one of the Milwaukee-matics, where both end faces are milled and internal bores are machined. The critical bell diameter is rough milled.

The next stop is one of the headchangers, where various hole patterns are drilled. The transmission case is then routed to an in-line refixturing station. Clamp pressure is relieved manually before the part is sent to the precision boring machine. This manual operation is necessary because the bell diameter must be held to a tolerance of only 0.004

in. Mainshaft and countershaft bores must be held to 0.001 in. Experience has shown that without such a stress-relieving operation, an out-of-round condition of 0.002 in. would occur—holding the size tolerance would be almost impossible.

When the finish boring operating is complete, the part is routed back for manual unloading. Workpieces are inspected off line using a Bendix Cordax coordinate measuring machine. Equipped with a Hewlett-Packard 9835 computer, the machine can check all dimensions on any workpiece in 80 minutes or less. Compound boxes are machined in a similar way. Two load stations are used for transmission cases and one for compound boxes, although the system can handle any mix of parts in random fashion. Only four different fixtures are required in the entire system.

One of the more interesting features of the system is the four-way shuttle mechanisms employed at all machines in the system. They serve to facilitate loading and unloading of pallets to the machines and provide substantial buffer storage at each cell. In this way, machine waiting time is all but eliminated.

The central control of the Mack FMS is handled by a DEC 11-44 minicomputer. A second DEC computer is used for backup. Communications capability to an IBM 370 is possible but rarely needed at this early stage of system implementation. Although the system is capable of producing 144 transmissions per day, current requirements are only about 50 per day. In this low-volume mode, the central FMS DEC computer is not used to schedule part mix and load information. The operator simply loads the parts that are required, then uses a nearby terminal to inform the DEC computer. Only nine programs are required to run the complete family of parts and each part number is cast into the workpiece for easy identification.

In the future, as production increases and service part requirements are transferred to the FMS, the two available positions in the FMS may be filled with additional Milwaukee-matics. Each piece of equipment in Mack's FMS has its own coolant and chip handling system. And, although it's doubtful that anything but aluminum will be run on the system, this increased flexibility could come in handy if cast iron parts must be processed in the future.

(continued)

LTV AEROSPACE AND DEFENSE

Vought Aero Products Division fires up one of the most advanced flexible manufacturing systems (FMS) systems to date: Over 500 complex parts for the B-1 bomber are machined using the latest flexible technology. Over 2000 machined parts are needed to assemble aft and aft-intermediate fuselage sections of the B-1 bomber. Vought was selected by Rockwell International to build the sections. The 2000 parts would require over 900,000 hours of machine tool time using conventional techniques. Vought's goal was to machine in-house the greatest number of parts at the least capital investment with an acceptable return. A flexible manufacturing system proved to be the best solution.

Exhaustive studies at the Dallas-based company showed that a flexible system with eight automated machining centers could perform the work of five hydrotels, eight milling machines, four three-axis profilers, and one standard machining center. Further studies showed that such a flexible system could perform the equivalent of over 200,000 hours of conventional machining in only about 70,000 flexible machining hours. An FMS could produce a $25 million cost avoidance in the B-1 bomber project and generate a seven-year depreciation write-off.

Vought refers to its installation as a flexible machining cell (FMC) because it represents one of a number of such cells the company hopes to install in the coming years. By linking so-called cells for flexible fabrication, painting, assembly, and other processes, Vought seeks to create a highly flexible, computer-integrated factory.

Over 500 parts were selected for machining in Vought's flexible system; 95% of the parts are aluminum, 5% are ferrous. Unusual in one respect, the Vought FMS consists only of machining centers—no headchangers, no special equipment of any kind.

The system is built around eight Cincinnati Milacron model 20-HC-1100 computer numerical control (CNC) machining centers, equipped with Acramatic 900 controls. The 30-horsepower, horizontal spindle machines can handle a prismatic work area of up to 32 x 32 x 36 in. Storage capacity is 90 tools and maximum workpiece weight is 5000 lbs. At this time, several machining centers have yet to be integrated into the system, but when fully operational, all eight machines will be available for production. A ninth machining center may be installed in the future for CNC program tryout and verification.

Two model 34-04P coordinate measuring machines manufactured by Digital Electronic Automation, Inc., play important roles in the Vought FMS. Every workpiece is initially checked 100%. Each controlled by a DEC PDP 11-23 computer, the coordinate measuring machines have repeatability of two-tenths.

Also part of the system is an automated parts washer. After every machining operation, the part is routed to the washer to flush chips and prepare the workpiece for inspection. Custom-built with 360-deg rollover, the washer is designed for complete untended operation. Also part of the system is a manual wash area where parts can be flushed by hand during planned maintenance of the automated washer.

The load and unload areas of this state-of-the-art FMS are unique. The loading function is performed at workstations located around the pallet carousels. This arrangement provides significant buffer storage for both finished and unfinished workpieces. More important, it allows enough storage of unfinished workpieces so that the FMS can run untended for long periods of time.

Because so many different parts are to be machined in the system, a separate area is set aside for fixture build up and tear down. Other stations in the system are reserved for reviewing workpiece quality and manual inspection, if required.

The automated material handling system currently is provided by four Robocarrier automated guided vehicles (AGVs) by Eaton-Kenway. The battery-powered Robocarriers automatically move to a charging station when power runs low.

One of the most advanced features of the Vought FMS is the provision made for automatic tool delivery to the machining centers. In the future, tools may be moved automatically from a computerized tool setup area to machines using Robocarriers. A cart-conveyed robot or machine-resident tool arm will be used to exchange tools. The network of floor-embedded wires needed to guide the Robocarriers to each tool changer is already in place.

Also somewhat unusual is the dual-purpose chip control system used by the Vought FMS. The central chip recovery and coolant distribution system, manufactured by Henry Filters, Inc., automatically segregates ferrous and nonferrous chips using flume diverters. Chips are not contaminated, so salvage value is higher.

(continued)

Case Study continued

(continued from page 323)

The Vought FMS runs up to 540 different parts, requiring nearly 1200 CNC part programs and 900 verification programs for part geometry measurement. Most parts are machined from plate stock and require an average of three passes through the system. Tooling holes are usually machined in the initial pass—parts are refixtured for the finishing passes. Seven different kinds of fixture types are used, including modular and vacuum fixtures.

The machining sequence starts at the fixture buildup area, where an operator manually mounts the appropriate riser and fixture on a standard pallet. The build schedule is displayed at a terminal near the station. Color graphics and text are used to assist the operator in building the fixture. Fixtures ready for loading are automatically routed to one of the two load-and-unload carousels, where operators load the appropriate workpiece blank.

After each machining sequence, the part is automatically washed, then routed to one of the coordinate measuring machines for inspection. If a part is shown to be out of tolerance, it is automatically routed to the second coordinate measuring machine for reinspection. If the second inspection confirms the out-of-tolerance condition, the part may be automatically routed to a material review station for manual inspection. If the suspect dimension is proved faulty, the system signals the central computer to put a hold on the suspect machining center used to produce the bad dimension. Simultaneously, terminals in the control room instruct operating personnel which tools should be checked out for problems.

After inspection, parts are routed back to the load and unload carousels. Fixtures are either reloaded or sent back to the fixture stations for teardown.

The control system of Vought's FMS is driven by a DEC 11-44 host computer with a DEC 11-70 as backup. The host computer downloads CNC programs, provides overall system control, and communicates with an IBM 3081 business computer. A networked pair of DEC 11-24 computers are used to control the Robocarrier system and the coordinate measuring machines. In addition, an Allen-Bradley Data Highway Network of programmable controllers is also employed.

Advanced software allows dynamic rescheduling using a 20-day window. Using simulation techniques, the work order schedule is daily updated to maximize throughput.

The Vought FMS is advanced in many other respects. One of the most important features of the system is its use of coordinate measuring machines. The Vought system uses part quality data to generate corrective action in real time. Although the loop is not completely closed yet, it represents a major step forward in automated quality control in FMS.

The number of parts to be processed on the system and the vast diversity of part configurations sets the Vought system apart from every other FMS. It proves that flexible machining technologies are applicable beyond the narrow scope of similar parts within a tightly defined family.

Major Topics Covered

- Personal Computers in Manufacturing
- Statistical Process Control
- Computer-Assisted Process Planning
- Simulation
- *Case Study: Weyerhaeuser Company*
- *Case Study: GSE, Inc.*
- *Case Study: Caere Corporation*
- *Case Study: IBM*

Chapter Ten

Automated Assembly

Automated Guided Vehicles

CAD/CAM

Numerical Control

Industrial Robots

Lasers in Manufacturing

Programmable Logic Controllers

Flexible Manufacturing

Computer Integrated Manufacturing

Other Related Technologies

PERSONAL COMPUTERS IN MANUFACTURING

The **personal computer** (**PC**) has become a widely used tool in manufacturing. The PC has had a great impact in the area of computer-aided design/computer-aided manufacturing (CAD/CAM). It is also widely used in offices for word processing and other office support functions. Personal computers are also used as tools in planning, controlling, and scheduling the production of products and in controlling the quality of the products.

Personal computers are among the most widely used tools in the modern manufacturing plant. They are used in manufacturing management and human resource planning, marketing, strategic planning, and finance. They are a key ingredient in factory automation systems and can be found in materials handling, assembly, inspection/testing, and materials processing applications. In the area of manufacturing planning control, they are used in a variety of shop floor, materials handling, scheduling, quality process, and facility planning settings. In the area of product/process, they can be found in design, drafting, analysis, simulation, and documentation applications.

No other tool used in manufacturing is used in all of these various manufacturing situations. There are many specific examples of how personal computers are being used in manufacturing. This section focuses on the way personal computers are used with computer numerical control (CNC) machines, robots, in computer-integrated manufacturing (CIM) systems, for production scheduling, quality control, cost estimating, process planning, and purchasing.

Personal Computers and CNC

Computer numerical control is an automated approach to the control of manufacturing machines such as mills, drills, and lathes. Prior to the development of the PC, numerical control (NC) systems had three basic components: the input medium, controller/reader unit, and machine tool.

In the early days of NC, the input medium was a paper tape that was punched to code the program for controlling the machine tool. Paper tape, due to its fragility and the difficulty in correcting errors, was replaced by magnetized plastic tape. Once the input medium was coded, it was fed through the controller, which read the instructions and interpreted them for the machine. The controller then fed the instructions to the electric motors that actually powered the machine tool.

Numerical control represented a significant improvement over manual control of machine tools. However, it was not without its problems. Problems commonly associated with traditional NC include (a) programming errors in preparing the input medium, (b) punched tape is fragile and incapable of repeated use without wearing out and breaking, (c) it is difficult to correct an error punched into paper tape, (d) tape readers are traditionally unreliable, and (e) the conventional NC controller unit is hardwired and inflexible. All of these problems have been solved through the use of the microcomputer.

In a conventional NC system, the tape must be fed through the reader and into the controller every time a particular programmed function is to be carried out by the machine. This means that the tape must be fed through and run repeatedly for each successive program execution. However, with CNC, the program can be fed into a microcomputer just once and stored. Then, each time the programmed instructions are to be fed to the machine, they are simply called up from memory and forwarded. In addition, using a microcomputer as the controller allows for greater flexibility. Since microcomputers are not hardwired like traditional NC controllers, they can be reprogrammed.

Computer numerical control using a microcomputer offers a number of advantages. The most important of these are (1) the ability of a microcomputer to accept the program and store it, rather than having to continually refeed it, as in traditional NC systems; (2) on-site program editing; (3) automatic conversion from among measurement systems (for example, converting from American standard to metric measurement); and (4) greater flexibility.

PCs and Robotics

Industrial robots are widely used in manufacturing. They typically perform repetitive, boring, or dangerous tasks traditionally performed by human beings. Such tasks include materials handling, machine loading, welding, spray painting, inspection, and assembly. An industrial robot is a computer-controlled electromechanical device. The type of computer most often used to control robots is the PC.

There are four different ways to program robots: the manual method, the walk-through method, the lead-through method, and off-line programming. Of these, the most efficient is off-line programming for CNC. A program for guiding the robot through the various required motions is written using a PC and then stored. With off-line programming, unlike the other three methods, the robot does not have to be stopped and taken away from its work to be programmed. With off-line programming, the program can be written while the robot is working on another operation.

The programs for guiding a robot through the desired motions can be written on the same PC used in designing the part. Figure 10–1 is an industrial robot programmed using a PC.

PCs and CIM

Computer-aided design and drafting (CADD), CNC, and robotics are all individual developments and concepts that represent steps up the ladder toward the eventual realization of the completely automated factory. Computer-integrated manufacturing (CIM) represents the total integration of these various individual concepts and several others. The CIM system will be the major component of the totally automated factory of the future. In such a system, CADD, CNC, robotics, and materials handling will all be integrated. Overall control will be provided by computers. Powerful PCs will be able to serve this purpose.

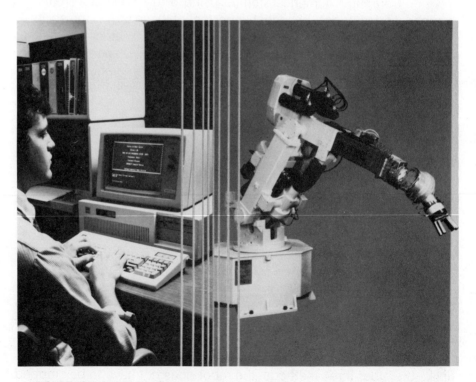

FIGURE 10–1 Industrial robot programmed using a PC. *Courtesy of Cincinnati Milacron.*

In a CIM system, individual intelligent PCs are used within the CADD component, the CNC component, the robotics component, and the materials handling component. In such a system, there is instant continuous communication among all components.

PCs in Production Scheduling

Production scheduling is a common manufacturing task that can be simplified with the aid of a microcomputer. In scheduling production jobs, there can be many possible scheduling sequences. For example, the number of possible sequences for running X jobs over Y machines can be calculated as X factorial to the Y power. To run just four jobs across five machines would be

$$1 \times 2 \times 3 \times 4 = 24^5.$$

This is an extraordinarily large number of possible sequences from which to select the best sequence. To do so manually, or even with a handheld calculator, would be a difficult task. However, performing such tasks is one of the strengths of PCs. This characteristic has made the PC an invaluable aid in production scheduling.

PCs in Quality Control

Quality control and metrology are parts of manufacturing. Two major elements of quality control are inspection and testing. Personal computers can be used to enhance both processes. Inspection and testing have traditionally been accomplished using a variety of gauges and special devices. These processes can be accomplished faster and more accurately using a PC and various sensor devices. Using a PC and contact as well as noncontact sensors, all of the various characteristics of a part (i.e., shape, length, width, thickness, feature locations, parallelism, perpendicularity, flatness, circularity, profile, concentricity, symmetry, etc.) can be tested much faster and more completely than with manual methods. The speed and storage capabilities of the PC allow 100% testing, rather than the sampling methods used with manual techniques.

PCs in Cost Estimating

Cost estimation is a time-consuming, difficult, error-prone task when undertaken using traditional manual techniques. The process can be simplified and made more efficient with the aid of a PC. The task of estimators is to determine how much it will cost to produce a product. This means they must estimate all costs, including labor, material, overhead, and hidden costs so that the profit margin can be added.

Accurate cost estimating is essential in manufacturing. There are PC-based cost-estimating software packages widely available that contain all of the estimating parameters needed in a manufacturing setting. With such programs, an estimator need only enter the appropriate information as prompted by the PC. The PC performs all calculations, research, and retrieval tasks required. The results of every estimate performed are categorized, stored, and updated as needed. With such a program, anyone who is able to respond to the prompts can produce an accurate estimate in a short time. Many CAD/CAM systems now have cost estimation software that will run on a PC (Figure 10–2).

PCs in Process Planning

Process planning, like estimation, is a time-consuming, difficult, and error-prone task when undertaken using traditional manual meth-

FIGURE 10–2 PC-based CAD/CAM system. *Courtesy of PMX, Inc.*

ods. However, the process can be improved with the aid of a PC. Process planning involves layout, tool design, and setup.

Layout involves developing a plan for meeting the tolerance, finish, feature, and diminishing requirements of a job. Most errors in layout are caused by insufficient time to perform all of the calculations necessary to determine the optimum cycle time, mathematical mistakes, insufficient allowances for secondary operations, and mistakes in reading drawings.

Tool design involves a variety of mathematical calculations, looking up figures in tables, and making drawings, all of which are error-prone tasks when accomplished manually. Setup, done manually, is also an error-prone process. It involves preparing specific detailed instructions. Such instructions must cover gauge point locations, the tools needed, and the physical identification of all fixtures, gears, cams, and attachments.

Personal computer-based interactive graphic software is available for layout, tool design, and setup. In each case, the PC and the program do the work. The operator simply responds to prompts. The advantage of the graphic capability of the PC is that it allows manufacturing personnel to actually see the impact of the responses they give to prompts and to react accordingly (Figure 10–3).

PCs in Purchasing

Purchasing is an important element in the overall production process. All of the materials, spare parts, and components needed in manufacturing must be purchased at some point. The more efficient its purchasing operation, the more efficient a manufacturing firm will be and the less costly in-house inventory it will have to maintain.

The PC can streamline the purchasing process. Several vendors produce software that allows manufacturing firms to network their purchasing office with its most frequent suppliers. In this way, items can be selected and ordered by the PC. This is more efficient than the old method of searching through volumes of catalogs, filling out requisitions, and preparing purchase orders. It can even eliminate the step of calling several companies for quotes and bids. Within a PC network, the desired item can be called up, displayed on the screen along with pertinent data, and priced. The same process can be accomplished for several vendors of the same item. Once an order is placed by a manufacturer, the vendor's computer processes it, updates its own inventory, prepares the shipment, and writes the invoice.

FIGURE 10–3 PC-based CAD/CAM system. *Courtesy of CADKEY, Inc.*

STATISTICAL PROCESS CONTROL

A variety of processes are used in manufacturing to produce products. **Statistical process control (SPC)** is a modern approach to making decisions concerning the processes used to produce products. It applies statistical methods in making decisions that might lead to either overcontrol or undercontrol of a given process.

Although SPC was developed over 60 years ago, it did not achieve wide-scale use until the computer became a commonly used tool in manufacturing. Computers make it easy for manufacturing personnel to apply the laws of probability and statistical methods when making decisions concerning manufacturing processes. Statistical process control is the application of the laws of probability and statistical methods to decision making in manufacturing.

How SPC Works

Manufacturing processes are planned around a set of specifications. Variations from planned processes are of two types. The first type of process variation is known as common or system variation. The second type of process variation is known as special or assignable variation.

Variations that result from common causes will form a pattern that can be measured and described on a distribution graph. Variation from common causes is predictable and can be thought of as controlled variations. Factors that cause variations that cannot be described by distribution of the process output are referred to as special. Special causes are unpredictable and must be corrected quickly. Processes with uncorrected special variations are said to be out of statistical control.

Rationale for SPC

Traditional methods of quality control involve inspecting a random number of produced parts and using the information collected to make judgments about the quality of all products within the lot. This sampling approach to quality control led to widespread inefficiency and waste in manufacturing worldwide. In the 1970s, countries such as Japan and the Federal Republic of Germany adopted SPC as a means of becoming more competitive in the international market place, and it worked. This was one factor that led to the decline in marketshare experienced by American manufacturing firms during the 1970s. It was not until the 1980s that American manufacturing firms began to make progress toward regaining marketshare in the international market place by adopting SPC.

The continual improvement of both quality and productivity is the goal of SPC and the rationale for the adoption of SPC. Manufacturing managers should use SPC to continually identify and correct system faults as well as the sporadic or special variations that sometimes crop up in a manufacturing process.

A manufacturing process that is under SPC is performing at its best. This does not necessarily mean that the process cannot be improved. But a system operating under SPC cannot be improved without changing the system. Having processes under SPC is important because it ensures predictability, one of the keys to success. When manufacturing managers are able to accurately predict the performance of production processes, scheduling, inventory management, and maintenance management are easier to accomplish. Statistical process control also makes it easier for manufacturing managers to evaluate the capabilities of manufacturing processes.

Because processes running under SPC are stable and predictable, it is easy for manufacturing managers to take measurements that are consistent and reliable. Data collected from such measurements can be used to accurately extrapolate the future performance of a process with a high degree of reliability. Processes that are not under SPC are too erratic and unpredictable to allow for the consistent measures needed to predict future performance.

COMPUTER-AIDED PROCESS PLANNING

Once a part is designed, the processes used to produce it must be planned. In a traditional manufacturing setting, a wide gap exists between the design and manufacturing components. Productivity and quality can be improved when the design component works with the manufacturing component from the beginning of the design through completion of production of the part. **Computer-aided process planning (CAPP)** represents a major step toward bridging the gap between the design and manufacturing components.

Once a product has been designed, there is a wide variety of sequences that can be used within the various processes required to produce the part. In fact, the number of combinations of sequences is usually so large that, prior to computers, process planning was really just a guessing game. However, with the advent of CAPP, manufacturing personnel can easily determine the optimum sequence of operations for producing a part. Computer-aided process planning systems are expert computer systems that collect and store all known information about a specific manufacturing setting, as well as numerous general manufacturing and engineering principles. They then use this information to determine the optimum plan for producing a given part. Such a plan will specify the machines to be used in producing the part, the sequence of operations, the tooling to be used, optimum speed and feed settings for the tools, and any other data needed to produce the part. Two key concepts in CAPP are CAD and group technology.

How CAPP Works

With CAPP, the part is designed on a CAD/CAM system. The mathematical model that describes the part is transferred electronically from the CAD/CAM system to the CAPP system. Using information stored in the CAPP system, the computer matches the characteristics of the part to the machines and processes available on the shop floor. The CAPP system prints out the process sheets and routing sheets that make up the process plan.

Two types of CAPP systems are available: the **variant system** and the **generative system**. The variant system is currently the more widely used of the two. The variant system is based on the concept of group technology. It develops a process plan for a new part based on the classification and coding of other similar parts that have already been produced. Variant planning involves identifying in the database the part produced in the past that is most similar to the new part to be produced. The process plan for the previously produced part is modified appropriately to produce a plan for the new part.

The generative process planning system starts from scratch each time. Such systems have a database that contains a wide variety of manufacturing specifications, standards, logic, and the capabilities of available machines and equipment. The part description and specifications are entered into the CAPP system. The system then develops the optimum plan for producing the part.

Computer-aided process planning systems can be implemented as stand-alone systems or within a CIM environment. In fact, CAPP can provide the vital link between CADD and CAM that is still the Achilles heel of CIM.

SIMULATION

Simulation is an important concept in the modern manufacturing plant. Simulation is a method of using mathematical models of real systems to test or predict the actual performance of those systems. Simulation is another manufacturing area in which the computer plays a vital role. Through simulation, manufacturing personnel can test a design, analyze a procedure, or assess the performance of a process before implementing the real thing. A simulation capability in a manufacturing setting can substantially decrease wastes and the amount of time required to produce a product or implement a process. There are a number of successful computer simulation software packages currently on the market (Figure 10–4).

Simulation Steps

The Society of Manufacturing Engineers in its book *Automation Encyclopedia*, lists ten steps in the development of a simulation process:

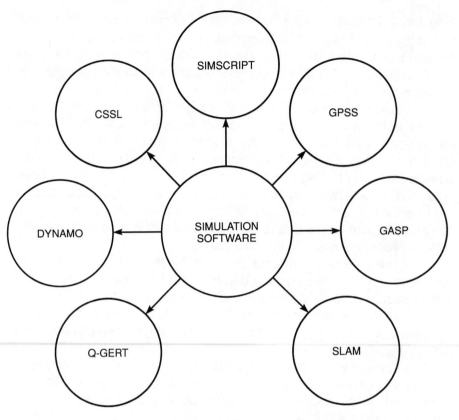

FIGURE 10–4 Computer simulation software packages.

1. Problem formulation: definition of the problem to be studied, including a statement of the problem-solving objective
2. Model building: abstraction of the system into mathematical-logical relationships in accordance with the problem
3. Data acquisition: identification, specification, and collection of data
4. Modes translation: preparation of the model for computer processing
5. Verification: process of verifying that the computer program processes as intended
6. Validation: establishing that the desired accuracy or correspondence exists between the simulation and the real system
7. Strategic and tactical planning: establishing the experimental conditions for using the model
8. Experimentation: execution of the simulation model to obtain output values

9. Analysis of results: analyzing the simulation outputs to draw inferences and make recommendations to resolve the problem
10. Implementation and documentation: implementing decisions resulting from the simulation and documenting the model and its use

Applications of Simulation

There are a wide variety of applications of simulation in manufacturing. A simulation model might be developed for a design so that the performance of the design can be assessed before the product is actually produced. A simulation model might be developed for a process or procedure so that its performance can be assessed before it is implemented. A simulation model of a robot in a workcell might be developed to assess the performance of the robot in the cell. A simulation model of a system of automated guided vehicles might be developed to ensure that the vehicles behave according to plan. Regardless of the way simulation is used, its goal is to allow manufacturing personnel to assess and predict how a design, procedure, or process will work prior to implementation.

KEY TERMS

Personal computer (PC)
Statistical process control (SPC)
Computer-aided process
       planning (CAPP)

Variant CAPP
Generative CAPP
Simulation

Chapter Ten REVIEW

1. Explain how the personal computer (PC) is used in a CNC system.
2. Explain how the PC is used in a robotics system.
3. Explain how the PC is used in a CIM system.
4. Explain how the PC is used in
       production scheduling
       quality control
       Purchasing
5. Define statistical process control (SPC).
6. Define CAPP.
7. Define simulation.
8. Explain the difference between variant and generative CAPP.
9. List the ten steps in simulation.
10. Explain the difference between common and special causes in SPC.

These case studies were provided by the Society of Manufacturing Engineers (SME). They are excerpted from the SME's *Manufacturing Insights®* series of videotapes. These case studies give students actual examples of the way the advanced manufacturing technologies covered in this chapter are being applied in "real-world" manufacturing settings.

As you read each case study, relate the examples cited to the material presented in the text of the chapter. This combination of textbook information and real-world examples will be particularly valuable to your understanding of advanced manufacturing technologies.

WEYERHAEUSER COMPANY

At Weyerhaeuser Company, the personal computer tackles the tasks of process control and production monitoring. Known worldwide for their highly diversified line of paper and wood products, including indoor paneling, corrugated cardboard packaging, and diapers, Weyerhauser has established itself as a leader in resolving manufacturing automation problems with the personal computer (PC).

At Weyerhaeuser's Marshfield, Wisconsin facility, Timblend brand particleboard and architectural wood doors are the principal products. In the particleboard manufacturing area, over 200,000 square feet of board is produced each day.

The process begins by taking low-cost wood scrap and grinding it into fine particles. Three sifters separate the wood particles into varying sizes to be incorporated into particleboard products with differing strengths and surface qualities. The particles are mixed with a high-temperature curing glue and formed into mats on large steel plates referred to as cauls.

In batches of up to ten cauls at a time, the particleboard mats are hot-pressed into panels. A thickness monitoring system utilizing three ultrasonic sensing gauges is used to control panel thickness. When a particleboard panel passes through the sensing units, six measuring heads emit small high-frequency sound waves that bounce off the board's upper and lower surfaces. The resulting distance information is then digitized by the gauges, and sent to an IBM 5531 PC/XT for thickness analysis and display.

With a color graphics monitor, the operator can view the maximum, minimum, and average thickness of each pressload of 10 boards, and intervals of 20 pressloads at a time. Another display shows multiple longitudinal cross sections of each board as it is being measured—information that the operator can use to determine whether the board's thickness is consistent throughout the panel. Other displays are used for inputting target thickness, production rate, and pressload number.

Using the thickness monitoring system ensures that the hot press platens are kept in excellent mechanical condition. It also keeps these wide expanses of steel, some as large as 8 ft across by 20 ft long, free from any deviations that could be easily passed on to the pressed panels. By maintaining proper board thickness, less sanding is required to meet the particleboard's final size. A cost-saving two-stage sanding operation brings the panels to within 1/2000 in. from their finished thickness.

In another building, the PC keeps track of Weyerhaeuser's line of architectural doors. Bar code labels applied to individual doors are scanned at key locations throughout the processing operations. The results of the scanning operations are sent back to the PC, which can (a) assign sorting locations to the different orders, (b) determine when to begin the processing of new orders, (c) print new bar code labels for the doors, (d) sort doors to proper locations, and (e) identify when an order is complete.

At Weyerhaeuser, innovative use of the PC has improved quality control, reduced costs, and added the advantage of knowing where and how much of the product is in process, so that Weyerhaeuser's First Choice quality can be delivered to the customer on time.

GSE, INC.

At GSE, Inc., the personal computer (PC) demonstrates the link between computer-aided design (CAD) and computer-aided manufacturing. GSE, Inc. in Farmington Hills, Michigan, is a world leader in the design and manufacture of physical measurement systems. In its 36,000 square foot facility, GSE designs and fabricates a complete line of torque sensors, load cells, special purpose transducers, measuring instruments, and control systems. The bulk of GSE's work involves accurate and efficient metal removal. Today, this task has been greatly simplified

(continued)

with the addition of an elaborate PC-based numerical control (NC) programming and CAD system.

One of GSE's current projects is to design and machine a biaxial load cell assembly from a hardened alloy tool steel. The completed assembly is used to test newly formed automotive tires for uniformity. A pair of the load cells are connected to a road wheel, which becomes the driving mechanism for the mounted tire. As the tire rotates, the load cells measure the biaxial forces exerted on the tire by the road wheel and identify any areas of imbalance.

With the aid of Cadkey, a CAD software package developed by Microcontrol Systems Inc., the part is designed, altered, and redesigned with ease on a Sperry IT personal computer. Having built an extensive library of design elements such as bending beams, torque tubes, hollow squares, and direct stress and load zone sections, GSE's CAD system is well suited to creating any part type from the variety of stored geometries.

After all the design modifications have been made, the Cadkey package generates an initial graphics exchange specification (IGES) file totally defining all of the geometric entities involved in the load assembly's design. This information is then translated into NC programming code by another PC software package, XL/NC, developed by PMX, Inc. In a matter of minutes, the XL/NC software automatically resequences the part geometries in an order for machining and then generates a corresponding tool path. With the seven different resequencing options offered by the XL/NC software, such as sequencing the NC file according to coincident end points or converting IGES points and full circles into base points to be used as center points for drilling, boring, tapping, and reaming operations, there is little more for the programmer to do than select the appropriate machine tools, feeds and speeds.

With the intended flexibility of the programming operation, the NC code can be transferred directly to the machining centers from the hard disk out of the PC or converted to punched tape and manually loaded onto the machines. Both methods allow the intricacies of a part's design to be stored easily for future machining jobs. Machining of the biaxial load cell is performed on a high-precision Hitachi-Seiki 4NE600 lathe and a four-axis Monarch WMC75 machining center.

The load cell is ready for delivery from GSE in just 3 days, a process that would have taken several weeks for a worker to manually design and program.

CAERE CORPORATION

At Caere Corporation, manufacturing resources planning (MRP) on a micro personal computer (PC) has nearly doubled projected output. Based in Los Gatos, California, Caere Corporation is a well-known manufacturer of bar code scanning and optical character reading equipment. With increased demand for their more than 200 models, Caere wanted to manage their growth with an efficient manufacturing control strategy. But because of the company's limited resources, the cost of implementing a mini or mainframe MRP system could not be justified.

In September 1983, Caere discovered a microcomputer-based MRPII system that was both affordable and offered the benefits of the larger systems. Using Max, The Production Manager, a MRPII software package developed by Micro MRP, Inc., Caere now maintains closed-loop, real-time control over every manufacturing operation.

The Max package consists of six basic modules:

1. bill of materials
2. inventory control
3. master scheduling
4. material requirements planning
5. purchasing control
6. shop floor control

Each buyer has a PC. When dealing with vendors on the phone or in person, the buyer can check current inventory, delivery schedules, and pricing, add new orders, and reschedule others. Purchase orders are entered into the master scheduling module, which defines the demand for the parts in each order based on the bill of materials. This module outlines the components that make up each product.

Next, a material requirements planning (MRP) explosion on the order generates a product tree consisting of all parts and assemblies, including information on the time it will take to fill the order and the lead

(continued)

time needed to match the subassemblies into the finished product. This information is then directed onto the shop floor, where the parts are pulled from stock.

Using their own PC scanner system, Caere records the quantity from a menucard and the part number from the container. The results of these transactions are entered directly into the MAX system as though the data had been manually typed in by keyboard. The MAX package immediately updates parts inventory and monitors the status of the order from assembly to shipment.

Caere's overall MRPII system is based on six IBM PC terminals connected by a Novell local area network. This network, combined with the MAX system has reduced inventory by nearly 26%, increased MRP explosion speed by 400%, reduced typing time for purchase orders by 50%, and given Caere 80% of the functionality of a mini MRPII system for only 15% of the cost.

IBM

At IBM, the personal computer (PC) serves as a twofold materials handling controller for a storage and retrieval system and an automated sorting machine. Based in Greencastle, Indiana, the IBM Central Distribution Center is responsible for the worldwide distribution of spare parts for PCs and other office equipment. With a direct customer order policy in effect since 1982, all domestic orders made through the Greencastle facility are delivered directly to the customers, without the interference of other distribution substations or parts centers. The operation handles up to 10,000 items per day, boasting a turnaround time of less than five days on 99% of all orders.

Each day orders are entered into a 3081 IBM host computer. The computer generates an extract file containing a unique sequence number for each record or item on order and a sort control field, such as an order number or customer number. Identification tags containing the item sequence numbers are then printed in conjunction with this file.

The extract file is first downloaded to the mini-stacker file server, an IBM PC/AT, which is linked to a network of industrial PCs and six Litton mini-stacker storage and retrieval machines.

From the file server, storage location information is sent to the corresponding PC workstations on each aisle of the Litton mini-stacker. These PCs supply the actual storage/pull locations to the Litton mini-stacker and display pick screens to the operators. The mini-stackers retrieve the different bins containing the parts specified by the pick instructions, and deliver them to the front of the stacker aisles. The bins are placed on rollers and lifted up to the operator for unloading.

Using the keyboard, the operator reports the items picked or unfilled. The parts are then tagged with the identification labels and sent in totes to the sorting machine.

In another area of the Greencastle facility, the sorting operation receives the picked items and completes the order-filling process. A Kosan Crisplant sorter is controlled by an IBM PC/XT, also known as the sorter PC. Using the sort control field from the extract file, the sorter PC assigns sort points to each order—the sort points are the actual sorter chutes of the Crisplant machine. If the sort control field is the order number, for example, then all items for a single order will be assigned to the same sort point.

At startup, using four induction stations, the sorter terminal operators key in the sequence number from the parts' identification tags and then place each item into a sorter tray. The sorter PC, having already determined the sort pattern to optimize throughput, delivers the part to the assigned chute. Once the part reaches the designated location, the tray tilts and the part slides into a tote box. The totes are then transferred to a packing area where they are prepared for shipment.

In the future, IBM plans to link the mini-stacker and the automated sorter by connecting demand printers to the PCs and bar coding the sequence numbers on the part identification tags. The result will be a fully integrated materials handling system exclusively controlled by industrial PCs.

Absolute system: A type of numerical control system in which all coordinates are dimensioned and programmed from a fixed or absolute zero point.

Access time: The length of time between the instant when a single unit of data is called from memory, and the moment when transmission is completed from the calling device.

Actuator: In robots, a device that converts electrical, hydraulic, or pneumatic energy into motion, i.e., cylinders, servo motors, rotary actuators.

Adaptive control: Machine control units for which fixed speeds and feeds are determined by feedback sensors rather than by being programmed.

Address: The location in a computer's memory of a word or block of data.

Addressability: The range of addressable points or device coordinates.

Addressable point: Any position specified in device coordinates.

Algorithm: A step-by-step procedure for achieving a given result by proceeding in a logical manner. Computer numerical control and computer programs are developed in this way.

Aliasing: The visual effects that occur when the detail of an image exceeds the resolution of the device space, e.g., a stairstep line on a raster display.

Allowed time: The leveled time with allowances for fatigue and delays added.

Alphanumeric display: A CRT display used to display text strings.

Analog computer: A computer working on the basis of a physical analogy (comparison of similarities) of the mathematical problem to be solved. The computer translates temperature, speed, voltage, and other physical variables into related electrical quantities and uses electrical equivalent circuits as an analog.

Annotation: The presence of textual descriptions on a display.

Artificial intelligence (AI): Built-in capability of a machine to improve its own operations; the capability of a machine to perform functions that are normally associated with human intelligence, such as learning, adapting, reasoning, self-correction, and automatic improvement.

ASCII (Pronounced "askey"): Abbreviation for "American Standard Code for Information Interchange;" ASCII is the standard code for expressing numbers, letters, and a variety of common typewriter symbols, such as period, comma, question mark, and carriage return, in a seven-bit format.

Assembly drawing: A CAD/CAM display that represents a major subdivision of a final product.

Assembly language: The most elementary language in which a human can program a computer.

Associative dimensioning: The updating of the respective dimensions of CAD/CAM display groups as the dimensions of their display entities change.

Asynchronous: A system where transmission of information between computers is under the control of interlocking (to and from) control signals rather than a direct function of clock cycles within a system.

Attribute: Any characteristic of a display item (color, linestyle, character font, etc.) or segment (visibility, detectability, etc.).

Automated assembly: Assembly by means of operations performed automatically by machines. A computer system normally monitors the production and quality levels of the assembly operations.

Automated materials handling (AMH): Handling and moving of materials using such automated devices as robots.

Automatic parts programming system (APPS): A system that allows an operator to manually trace a two-dimensional part to produce a numerical control part program.

Automated process planning (APP): Creation, with the assistance of a computer, of a process plan for parts in a given family.

Automatic programming tools: The combination of the APT language the APT computer program and the computer used for implementing the program.

Automated storage and retrieval (ASR): The use of a computer for the storage and retrieval of manufacturing and management data.

Automation: Converting a procedure, process, equipment, or system to automatic operation; the manufacturing process that produces a part by means of machines, mechanisms, and handling devices without the aid of human hands. The entire operation is controlled by electrical impulses from a computer or tape drive.

Axis: The imaginary line that runs through an object and about which a turning tendency exists; a rotary or translational (sliding) joint in the robot. Also called the degree of freedom.

B-spline: A mathematical representation of an arbitrary smooth curve.

Back annotation: The extracting of information from a completed printed circuit board to create a CAD/CAM display.

Back-up: Work that is copied onto a tape or disk to protect against losing data if the "user" tape or disk is damaged.

Batch processor: A computer that accepts programs in a continuous stream and executes them one by one, analogous to a batch manufacturing process.

Baud rate: The rate, in bits per second, at which data can be transmitted over a serial transmission line, such as a telephone line.

Benchmark: Any set of standards designed to verify performance specifications or to compare hardware or software produced by different manufacturers.

Binary code: Any system of representing data with zeros and ones only.

Bit: A single digit, 0 or 1, of a binary code.

Bit plane: The hardware used as a storage medium for image bit maps.

Blending: Mixing two or more lots of material to produce a homogeneous lot. Blends normally receive new identification and require testing.

Block: A fundamental unit of data on a disk or tape drive. A commonly used block size is 1024 bytes, although other block sizes can be defined by the computer system designer.

Block diagram: A graphic presentation of the paths along which data flow between various parts of a computer system.

Bomb-out: The complete failure of a computer routine, requiring a restart or reprogramming.

Boxing: A visibility test incorporated in clipping that uses a bounding box to test the relationship of an entire symbol to the clipping boundary.

Brinell hardness value: The hardness of a material indicated by the Brinell hardness testing machine. The value is convertible to other test machine hardness values such as Rockwell, Shore, Vickers, Knoop, and others.

Bronze: An alloy of copper and tin, often alloyed with a third element such as aluminum, silicon, manganese, or phosphorus.

Buffer: Temporary memory that speeds the flow of information between parts of a computer system by eliminating delays.

Bug: A minor computer malfunction usually caused by a programming error.

Bulk storage: Memory units (separate from the main memory of a computer) in which large amounts of data may be stored.

Burnish: The smoothing of part edges by rubbing actions in an especially made die.

Burr: The sharp ridge of metal nearly always remaining at the end of a sawing operation; the ridge remaining at the bottom side of a drilled hole.

Button device: A button used as a graphic input device.

Byte: A group of eight bits; a common measure of computer memory capacity.

Cast iron: An alloy containing the same basic elements as steel, but with a carbon content exceeding 1.7%.

Cathode: The negative electrode in an electric circuit. In an electrolysis process, the negative electrode is the cathode.

Cathode-ray tube (CRT): An electron tube in which electron beams projected onto its display surface excite the phosphor coating, producing luminous spots.

Central processing unit (CPU): The heart and brain of a computer, containing all the circuitry for controlling, timing, performing arithmetic, and addressing data for storage or transmission.

Centrifugal: The tensile force produced in a material due to rapid rotation about an axis. With respect to liquids, a centrifugal force causes the liquid to move to the periphery of a hollow spinning mold.

Ceramics: A material composed of compounds of metallic and nonmetallic elements existing as phases such as stone, brick, abrasives, glass, and concrete.

Character: An instance of a numeral, letter, or other linguistic, mathematical, or logical symbol.

Character font: A primitive attribute of text strings defining the style of the character set.

Character generator: A hardware device that accesses character patterns in a ROM and generates them at user-specific display surface positions.

Character plane: A primitive attribute of text strings defining the plane in which characters are generated.

Character size: A primitive attribute of text strings defining the size of characters in terms of the bounding box.

Chip: A very small silicon wafer containing an integrated circuit.

Chuck: A device for holding the workpiece during machining operations.

Clearance: The designed space left between two moving parts so that movement will occur as required within a specific temperature tolerance. The space must allow for lubrication as expansion and contraction occur.

Clipping: The process of determining which portion or portions of a display element lie outside the specified clip boundary and making them visible.

Coherence: A property used in raster scan that recognizes that adjacent pixels are likely to be similar in characteristics.

Cold heading: The cold deformation of a metal bar by impact loads for shaping, such as a bolt head.

COM: Computer output microfilm.

COM recorder: A display device for placing displays on microfilm.

Commercial diamond: Discolored diamonds used for industrial operations such as cutting and in hardness tester penetrators.

Comparator: A device that compares the proximity of a cursor to the vector currently being drawn.

Compatibility: The degree to which tapes, languages, programs, machines, and systems can be interchanged or can communicate.

Component: A CAD/CAM marker that has physical meaning, i.e., resistor, capacitor, switch.

Composite color: A color described in terms of its hue, whiteness, and blackness and encoded in a single video signal.

Composite video: A single video signal encoding color (red, green, and blue) data.

Compression strength: The maximum strength of a material while the material is under a pushing load as it collapses in some manner.

Computer: An electronic device that manipulates and processes data according to programmed instructions. Basic computers consist of input/output devices, a central processing unit, and memory devices.

Computer-aided design/computer-aided manufacturing (CAD/CAM): Using a computer to improve productivity in design and manufacturing. Includes CAD, CADD, CNC, robotics, and CIM.

Computer-aided design (CAD): The application of computers to aid or enhance the design process.

Computer-aided design and drafting (CADD): Using computers to aid the design process and in producing the documentation of a design.

Computer-aided engineering (CAE): Use of a computer, special software, and various peripheral devices to accomplish such engineering tasks as analysis and modeling.

Computer-aided manufacturing (CAM): The use of computers to aid in the management, control, and operations of manufacturing through direct or indirect interface with the human and physical resources of the organization.

Computer animation: The use of computer graphics to generate motion pictures.

Computer automated process planning (CAPP): Use of a computer to develop a detailed plan for the production of a part or assembly.

Computer graphics: A family of related technologies that permit digital computers to control, alter, and display pictures rather than only text or numbers.

Computer-integrated manufacturing (CIM): The total integration of such individual concepts as CAD, CNC, robotics, and materials handling into one large system.

Computer numerical control (CNC): Control of manufacturing machines and systems using a computer and programmed instructions.

Configuration: A group of machines, devices, parts, and other hardware that make up a system.

Connectivity: Data that describe how components of a system are connected.

Continuous path: A servo-driven robot that provides absolute control along an entire path of arm motion, but with certain restrictions in editing and ease of program change.

Contrast: The ration of the highest available intensity level to the lowest.

Controlled path: A servo-driven robot with a control system that specifies or commands the location and orientation of all robot axes.

Coolant: A liquid used during machining to cool the cutter and the metal being cut, to lubricate the cutting action, and to wash chips from the cutter zone.

Coordinate: The location of a point in terms of units from the specified origin.

Core memory: The main memory or random access memory of a computer.

Crash: A computer system failure that usually results in loss of data, and, therefore, loss of labor hours.

Critical items: Production items that require a lead time longer than the normal planning span time or items whose scarcity could limit production.

Critical path method (CPM): A special technique for scheduling resources to accomplish a job within all applicable constraints.

Cross hairs: Two intersecting perpendicular lines incorporated in a cursor, with the intersect being used to indicate desired device coordinates.

CRT display: A display device employing a cathode ray tube.

Cupola: The hot air furnace used in the production of cast iron.

Cursor: A recognizable display entity that can be moved about the display surface by a graphic input device to return either device coordinates or a pick stack.

Cutting speed: The recommended surface speed in feet per minute that a material turns during a cutting operation that results in maximum cutting efficiency for both tool and material.

Cybernetics: The science of control and communication systems. It encompasses (a) integration of communication, control, and systems theories; (b) development of systems engineering technology; and (c) practical applications at both the hardware and software levels.

Data processing: The performance of a systematical sequence of mathematical and/or logical operations that a computer performs on data.

Database: Any collection of information having predetermined and useful organization.

Database management system (DBMS): A software system for managing data and making such features as interrogation, maintenance, and analysis of data available to users.

Debug: Troubleshooting and correcting computer hardware and software problems that affect the operation or performance of the system or a program.

Decluttering: The selective erasure of display items when the display is too dense to easily discern details.

Deformation: Under a load, deformation first occurs as elastic, then inelastic as the yield is exceeded.

Delphi method: A forecasting method in which the opinions of experts are combined in a series of questionnaires. The results of each questionnaire are used to design the next questionnaire, so that convergence of the expert opinion is obtained.

Depreciation: The actual decline in the value of an asset due to exhaustion, wear and tear, or obsolescence.

Design rules checking (DRC):

Destructive testing: The physical destruction of a specimen or sample in determining the several mechanical properties of like materials and shapes.

Digital computer: A computer that solves problems through arithmetic operations on numbers composed of digits.

Digital vector generator: A device used with raster displays to interpolate the straightest possible pixel string between specified endpoints.

Digitize: To convert from graphic or analog form to representation by discrete values.

Digitizer: A data tablet that generates coordinate data from visual data through the use of a puck or stylus.

Dimensioning: The measuring of distances on a CAD/CAM display.

Direct numerical control (DNC): Machines controlled by a dedicated computer that stores numerous numerical control machine programs.

Direct-view storage tube (DVST): A CRT whose display is maintained by a continuous flood of electrons.

Directed beam: The technique used in calligraphic displays to produce vectors by having the electron beam stroke them in a selected order.

Disk: Flat, circular, magnetic recording device capable of storing large amounts of data.

Disk operating system (DOS): A computer program that interacts with the processor and the disk or diskette drive to control the flow of data.

Diskette: A very small, very inexpensive form of disk drive; sometimes called a "flexible disk" or "floppy disk."

Display: A collection of display items presented on the display surface.

Display device: An output device used to display computer-generated graphical data.

Display entity: A logical grouping of output primitives that forms a recognizable unit on the display surface.

Display file: A collection of display instructions assembled to create a display.

Display image: The portion of an image visible on the display surface at any one time.

Distributed numerical control (DNC): The use of a shared central computer for distribution of part programs and data to several remote machine tools.

Diversion of work: The separation of tasks into less complex subtasks, the basis for work specialization.

Dot matrix: A pattern of dots taken from a two-dimensional array.

Dot matrix printer: A plotter that produces displays in dot matrix form.

Dragging: The interactive mode technique of moving a display item by translating it along a path determined by a graphic input device.

Draw: The generation of a vector by creating a line segment from the current position to a specified endpoint, which becomes the new current position.

Drum plotter: A plotter whose display surface is a rotatable drum and whose plotting head can only move parallel to the drum's axis of rotation.

Ductility: The ability of a material to be plastically deformed without rupture as the stress increases beyond the yield strength and below the tensile.

Dumb terminal: A terminal that receives and displays data but is incapable of altering it.

Echo: The mode of a graphic input device that provides visual feedback to the operator, e.g., a cursor, text string, etc.

Elastic: The ability of a material to be temporarily deformed as the stress increases up to any value below the elastic limit or the yield strength.

Electron gun: The part of a CRT that focuses and emits the electron beam.

Electrostatic plotter: The part of a CRT which focuses and emits the electron beam.

Element: A basic graphical entity, i.e., a point, line segment, character, marker, or text string.

Emulation: The use of special programming techniques and machine features to allow a computing system to execute programs written for another system.

End-effector: The tool attached to the robot manipulator or arm that actually performs the work.

Endpoint: Either of the two points that mark the ends of a line segment.

Endpoint matching: The accuracy of the vector generator in drawing two or more vectors emanating from the same point.

Expansion card: A card on which additional printed circuits can be mounted.

Expediting: The rushing or facilitating of production orders that are needed in less than the normal lead time.

Fabrication: Processing of materials for desired modification of shape and properties.

Family of parts: A group of parts having similar topology or manufacturing characteristics.

Fatigue: The combination of a cyclic load and time on a material that ultimately leads to material failure, even though the stresses are far below the tensile strength.

Ferrous: Iron-based metals including wrought iron, steel, and cast iron.

Field: A part of a data record set aside for a specific element of data (see database).

File: A collection of data, in the form of records, that exists on a computer disk or tape.

Fill: To fill an area of the display surface bounded by vectors, e.g., with a solid color or a pattern of line segments.

Fillet: A radius at a change in mass of material.

Finite-element analysis: A method for detailed stress analysis of irregularly shaped structures, such as automobile or airplane frames, nuclear reactor nozzles, etc.

Finite-element model: A mathematical model of a continuous object that divides the object into an array of discrete elements for simulated structural analysis.

Firmware: Sets of computer instructions (programs) cast into ROM.

Fixing: A device needed to hold a workpiece in proper position for work performance.

Fixture: A device for holding the workpiece during fabrication or machining.

Flatbed plotter: A plotter with a flat display surface fully accessible by the plotting head.

Flexible automation: Multitask capability of robots; multipurpose, adaptable, and reprogrammable automation.

Flicker: A noticeable flashing of the display during each refresh, caused when the refresh interval exceeds the phosphor resistance.

Floating point arithmetic: Arithmetic performed with real numbers, such as 12.5, 3.1417, and 87.66, rather than integers, such as 1, 753, and 56.

Floppy disk: A storage device consisting of a flexible magnetic disk inside a protective plastic jacket.

Flowchart: A diagram that outlines the logical steps a computer program should take, facilitating program design or documentation.

Fonts: Repetitive patterns used to give meaning to a line such as hidden lines, centerlines, or phantom lines.

Force: The weight or pressure on a material that endeavors to deform it.

Function button: A button on a device that can operate in either momentary or latchable mode and whose value may be retained

Function key: A key on a function pad that causes execution of special program functions defined by the user.

Function pad: A graphic input device with user-programmable function keys.

Gauge blocks: Very accurate metal blocks that act as standards of measurements at a given temperature.

Geographical numerical control (GNC): A part programming system that provides effective tape generation by providing graphic displays of the part, the tool path, and the tools themselves.

Graphic input: Any inputs entered by a user through a graphic input device while in the interactive mode.

Graphic input device: The hardware that allows the user to enter data or choose a detectable display item.

Graphics: The visual presentation of data as a series of output primitive.

Graphics package: A series of software routines that provide the user access to the graphics hardware for generating a display.

Graphics processor: A controller that accesses the display list, interprets the display instructions, and passes coordinates to the vector generator.

Graphics tablet: A peripheral device for drawing on a computer graphics system.

Gravity feed: Supplying materials into a machine, workstation, or system by the force of gravity.

Grid: The uniformly spaced points in two or three dimensions within which an object may be defined.

Group classification code: Material classification technique that designates characteristics using successively lower order groups of code.

Group technology: A means of classifying parts into families on the basis of similarities.

Hard copy: Information printed on paper instead of displayed on a CRT.

Hardness: The resistance of a material to penetration.

Hardware: The assembly of electronic components that constitute the physical makeup of a data processing system.

Hatching: The filling of an area of the display surface bounded by vectors with a pattern of parallel line segments.

Heat treat: The heating and cooling of a metal to induce desirable mechanical properties.

Hexadecimal: Numbers written in base 16 rather than base 10; used to display binary data because the hexadecimal form is more compact, yet can be easily translated into binary numbers.

High-resolution graphics: Graphic terminals that show great detail because they are made up of many pixels.

Homogeneous coordinates: The coordinates used in matrix transformations to convert objects described in n space to a representation described in $n + 1$ space; i.e., x, y, z become wx, wy, wz, where w is the homogeneous factor.

Host computer: The independent computer to which a peripheral device, such as a terminal, plotter, or disk drive, is attached.

Hue: A characteristic of color that allows it to be named (i.e., red, yellow, green, blue), often defined by an angle representing its gradation.

Image: A view of an object.

Image bit map: A digital representation of a display image as a pattern of bits, where each bit maps to one or more pixels.

Image transformations: The application of a transformation function to an image after projection to the display area.

Indirect labor: Labor that does not add to the value of a product but must be performed as part of its manufacture.

Initial graphics exchange standard (IGES): A U.S. national standard for exchanging mechanical design data between CAD systems.

Ink-jet plotter: A plotter that uses electrostatic technology to first atomize a liquid ink and then control the number of droplets that are deposited on the plotting medium.

Input: Data or programs that go into a computer system.

Input device: Any device such as a digitizer or keyboard that allows the user to give information to the computer.

Input/output (I/O) device: A device, such as a disk drive, that can send and receive data from a computer.

Inspection by attributes: Inspection in which a part is accepted or rejected based on a single requirement or set of requirements.

Integrated circuit (IC): A microelectronic chip.

Intelligent robot: A robot that can control its behavior through the application of its sensing and recognition capabilities.

Intensity: A characteristic of color defining its percentage ranking on a scale from dark to light, specifying perceived brightness.

Intensity level: One of a discrete set of brightness levels attainable with a CRT.

Interactive graphics: A method that allows users to dynamically modify displays through the use of graphic input data.

Interactive processor: A personal computer that performs the user's instructions immediately on receiving them and displays the results immediately on completion.

Jig: A device for holding the workpiece during machining and then guiding the tool as it machines the material.

Job classification: The grouping of jobs on the basis of the functions performed, level of pay, job evaluation, historic groupings, collective bargaining, or any other criteria.

Job lot: A small number of a specific type of part or product that is produced at one time.

Job lot layout: A group of machines and equipment especially arranged to handle job lot production.

Joystick: A graphic input device that employs a movable lever to control the position of a cursor, for returning locator or pick information.

Kernel: A subset of routines from a graphics package that permits construction of elementary displays.

Keyboard device: A graphic input device that allows the user to enter characters or other key-driven values.

Keyway: A machined groove in a part, for example, a shaft, that holds the key for connecting the shaft to the driving gear. The keyway may be rectangular in shape or semicircular such as the Woodruff.

Labor cost: That part of the cost of a product attributable to wages.

Laser plotter: A plotter that produces display images on photographic film, in raster or vector formats, using a laser.

Layer: The logical subdivisions of the data contained in a two-dimensional CAD/CAM display, such that the subdivisions may be viewed individually or overlaid and viewed in groups.

Lead time: The time normally required to perform a given activity.

Library: A collection of predefined symbols that may be placed on a CAD system drawing.

Light pen: A graphic input device that generates a hit detection when a pick is made while it is pointed at a detectable display item.

Limit switch: A switch that is actuated by some motion of a machine to alter the electrical circuit and limit or guide movement.

Line style: A primitive attribute of lines that defines whether they are to be solid or dashed and a possible dash pattern.

Line width: A primitive attribute that defines the thickness of a line segment.

Local area network (LAN): A type of network hardware that does not allow computers to be separated by physical distances of more than 1000 yards.

Locator device: A graphic input device, such as a joystick or data tablet, that uses a cursor to provide coordinate information.

Logic diagram: A drawing that indicates the interconnection of the individual logic elements of an electronic circuit.

M Function: Similar to G function except that the M functions control miscellaneous functions of the machine tool, such as turning on and off coolant or operating clamps.

Machine center: A group of similar machines that can all be grouped together for purposes of loading. (See work center.)

Machine tool: A powered machine used to form a part, typically by the action of a tool moving in relation to the workpiece to perform such tasks as turn, drilling, cutting, etc.

Macro: A sequence of CAD system commands contained in a text file or grouped on a menu key, which may be executed automatically.

Magnetic tape: Plastic or Mylar tape that is coated with magnetic material used to store information.

Magnetic tape storage: The recording of binary data on a magnetized tape.

Mainframe: An overused, ill-defined term that refers to the central processing unit of a large data processing system.

Maintenance: The collection of procedures necessary for retaining an item in or restoring it to a specified condition.

Manipulator: The arm of the robot; encompasses mechanical movement from the robot base through the wrist.

Manufacturing: A series of interrelated activities and operations involving the design, material selection, planning, production, quality assurance, management, and marketing of goods and products.

Manufacturing planning and control systems (MP&CS): Special systems, usually automated, used to set the limits or levels of manufacturing operations in the future and to control machines, materials, and processes in the present.

Manufacturing process: A series of tasks performed on material to convert it from the raw or semifinished state to a state of further completion and greater value.

Mapping function: A method of transforming an image definition expressed in one coordinate system to another.

Mass production: A method of high-quantity production characterized by a high degree of planning, specialization of equipment and labor, and integrated utilization of all productive factors.

Material requirements planning (MRP): Process for identifying the types and amounts of materials required for future manufacturing projects using bills of material, inventories, and master production schedules.

Materials handling: The movement of materials, parts, subassemblies, or assemblies either manually or through the use of powered equipment, robots, or systems.

Memory capacity: The number of actions that a robot can perform in a program.

Menu: A list of program execution options appearing on the display surface that prompts the user to choose one or more through the use of a graphic input device.

Menu keypad: An electronic tablet that contains individual blocks that may be programmed with single commands or groups of commands.

Microprocessor: A central processing unit contained on a single integrated circuit.

Mirroring: The creation of a mirror image of a display image.

Modeling system: A system that allows models to be defined and transformed using world coordinates.

Modeling transformation: A transformation that transforms the world coordinate system of a model to the default world coordinate system of a graphics package that is in effect immediately prior to a viewing operation.

Modulator-demodulator (modem): A device that permits a digital computer to transmit digital data over the analog circuits of local telephone lines.

Motion study: An analysis of the movements that occur in an operation for the purpose of eliminating wasted movement and establishing a better sequence.

Mouse: A device for controlling the cursor. It looks like a small box, with several buttons on the top, and is connected to the computer by a cable.

Nesting: In programming, the grouping of individual commands or operations into a single command that contains all of the various operations.

Networks: Communication lines connecting computers and computer-controlled systems.

Node: The intersection of two or more interconnections.

Nondestructive testing: The testing and inspection of a material workpiece or part to determine mechanical properties without harming the material, workpiece, or art.

Nonferrous: Metals that are not ferrous, even though iron may be a constituent in some alloys.

Numerical control (NC): A method of controlling manufacturing equipment and systems that accepts commands, data, and instructions in symbolic form as in input and converts this information into a physical output in operating machines.

Off-line: Equipment or devices in a data processing system that are not under the direct and immediate control of the central processing unit.

On-line: Equipment or devices in a data processing system that are under the direct and immediate control of the central processing unit.

Open shop: A CAD facility in which designers from a variety of departments use design stations on a part-time basis.

Operating system: The primary control program of a data processing system. Written in assembly language, the program controls the execution of programs within the central processing unit and controls the flow of data to and from the memory, and to and from all peripheral devices in the data processing system.

Optical scanner: A video camera tube incorporating an electron beam to scan an input image, sense the light emitted, and produce video signals.

Optimize: In numerical control or computer applications, the arrangement of instructions to obtain the best balance between operating efficiency and the use of hardware capacity.

Output: Data that come out of a computer.

Painting: A technique similar to inking, but used only on raster displays where line width and color may vary.

Pan: To translate horizontally.

Paper tape: One type of input medium used in numerical control.

Parallax: The apparent displacement of a display item between where the viewer perceives it and where a light pen is pointing.

Parallel operation: The simultaneous performance of several actions (usually of a similar nature).

Parallel transmission: A system of transmitting information wherein the characters of a word are transmitted simultaneously over separate lines or wires (compare to serial transmission).

Part: A single manufactured item used as a component in an assembly or subassembly.

Parts list: List of parts used in an assembly or subassembly.

Passive graphics: A method allowing no operator dynamic interaction with a display.

Passive mode: A setting that specifies a display console as usable for passive graphics.

Pattern fill: Repetitively using a user-defined pixel array to perform fill.

Payload: Maximum weight that can be carried at normal speed.

Pen plotter: A vector that draws with ordinary ballpoint pens and ink.

Peripheral (device): Any device distinct from the central processing unit, such as disk drives, cathode ray tubes, plotters, and graphic tablets.

Phosphor: The coating of the inside of a cathode ray tube.

Pick-and-place robot: A simple robot with 2 to 4 axes of motion and little or no trajectory control.

Pickling: The chemical cleaning of a metal such as the pickling of steel in a bath of diluted sulfuric acid.

Pixel: The discrete display element of a raster display, represented as a single point with a specified color or intensity level.

Pixel array: A rectangular matrix of pixels.

Plasma panel: A type of display device whose display surface consists of a matrix of gas-filled cells that can be turned on and off individually and that remain on until turned off.

Plotter: Any device that produces hard copy of graphic data. Types of plotters include vector (pen), electrostatic, and ink-jet plotters.

Point-to-point: A servo- or non-servo-driven robot with a control system for programming a series of points without regard for coordination of axes.

Pop-up menus: Instead of having to memorize hundreds of commands, the program permits standard system commands to appear, as needed, on little "cards" that float on the screen.

Port: A physical connection linking a processor to another device or circuit.

Powder metallurgy: The science of compacting and sintering powdered metals into specific shapes.

Printed circuit board: A board on which a pattern of printed traces and connections has been etched.

Process: Series of continuous actions with a system of levels divided into subactivities that are accomplished by executing one or more tasks.

Production: Changing the shape, composition, or combination of materials, parts, or subassemblies in manufacturing.

Production capacity: The highest output rate that can be maintained without changing the product specifications, product mix, worker effort, plant, and equipment.

Production rates: The quantity of production expressed by a period of time, i.e., per hour, per shift, per day, per week, etc. (also called production levels).

Program: A set of instructions written in a data processing language that defines a task to be performed by a computer.

Prompt: Any action of the display console that indicates an operator reaction is needed.

Protocol: A set of rules governing the exchange of data between devices in a data processing system.

Puck: A handheld device with a transparent portion containing cross hairs that is used for inputting coordinate data from a data tablet through the use of programmable buttons.

Quality assurance (QA): A broad term that includes both quality control and quality engineering.

Quality control: The establishment of acceptable limits in size, weight, finish, etc. for products or services; the series of inspections performed on a workpiece or part from beginning of manufacturing to the end to assure those limits.

Random access memory (RAM): An array of semiconductor devices for temporary storage of data during the computation process.

Raster: A rectangular matrix of pixels.

Raster display: A CRT display whose display surface is covered by a raster and that generates displays using raster scan techniques.

Raster plotter: A plotter that produces displays in dot matrix form.

Raw material: Unprocessed material.

Read: To query a graphic input device and await operator action.

Read-only memory (ROM): Similar to RAM, except that the computer user may only retrieve data, not change it.

Real time: Computer processing or control performed at the same time as the controlled operation or process is occurring; accessing data that represent actual values at that point in time.

Refresh: The process of repeatedly drawing a display on the display surface of a refresh tube.

Refresh cycle: One refresh of the display surface.

Refresh display: A display device employing a refresh tube that permits dynamics because of a high refresh rate.

Refresh cycle: One refresh of the display surface.

Register: A circuit that stores bits in a central processing unit.

Repaint: To refresh a display surface with an updated display.

Repeatability: The accuracy of an analog vector generator in minimizing the deviation from precise overlap when redrawing vectors.

Resolution: The precision of a CRT, measured as the number of pairs distinguishable across the display surface.

Robot: An electromechanical device that performs functions traditionally performed by humans.

Rotate: To transform a display or display item by revolving it around a specific axis.

Routing sheet: A form listing the sequence of operations to be used in producing a particular part or product.

Routing: The positioning of interconnections in a CAD/CAM display.

Rubber banding: A programming technique that allows lines drawn on a computer graphics display to be stretched as if they were made of rubber bands.

Saturation: A characteristic of color defining its percentage difference from a gray of the same blackness.

Saved: Work that is filed or stored on a disk drive, so that loss of power or machine failure will not destroy it.

Scale: To transform the size or shape of a display or display item by modifying the coordinate dimensions; the ratio of the actual dimensions of a model to the true dimensions of the subject represented.

Scan line: A horizontal line of pixels on a raster display that is swept by the electron beam during refresh.

Scanning pattern: The path followed by an exploring spot.

Scanning spot: The point of focus of the electron beam of an image digitizer on the input image.

Scissoring: The process of determining the portions of a display element that will not be visible.

Screen coordinate system: A coordinate system that represents the internal digital limits of the display device.

Scrolling: The vertical translation of text strings or graphics.

Sector: A subdivision of a block, usually the smallest unit of data that can be retrieved from disk or tape memory.

Serial transmission: The transmission of data over a single pathway.

Servo system: A control linking a system's input and output, which provides feedback on system performance to regulate the operation of the system.

Servo mechanism: An amplifying device that takes input from a low-energy source and directs an output requiring large quantities of energy.

Setup lead time: The time needed to prepare or setup a manufacturing process.

Shading: An image processing technique that indicates light sources in a three-dimensional image; the changes in sensitivity of the video camera tube of an image digitizer.

Shear strength: The maximum strength of a material under cutting or shearing stresses. The load is applied across the cross section of the material.

Shielding: The defining of an opaque viewport or window in which to display a menu, a title, or a message to the operator.

Signal: A logical linking of pins in a CAD/CAM display using interconnections.

Simulation: The design and operation of a model of a system in a manner analogous to the way the real system operates.

Soft copy: A copy of a display in video form, as on videotape.

Software: Programs, procedures, and associated documentation for data processing systems.

Solids modeling: The construction of a "solid" model of a part from mathematically defined solid primitives (i.e., cubes, cones, and spheres) or surface primitives (i.e., planes, spherical segments, and deformed surfaces).

Specification: A statement of the technical requirements of a material, a part, or a service, and of the procedure to be used to verify that the requirements are met.

Stand-alone workstation: A CAD workstation containing microprocessors and capable of independent operation without being connected to a host computer.

Storage: The process or location for holding data or instructions inside a memory device.

Storage tube: A CRT that maintains a display on the display surface without refresh.

Strain: The change in dimension of a material as a result of the stress that accompanies a load. The strain is usually understood to be linear.

Stress: The internal reaction in a material to an external force, weight, pressure, or load measured in pounds per square inch.

Stroke writing: A line drawing, as opposed to a raster scan.

Stylus: A device analogous to a pencil, which is used in conjunction with a data tablet to input coordinate information.

Subroutine: A named display item description contained in the display list, used to create multiple views of the item without repeating the display instructions.

Synchronous: An operation that takes place within a fixed time interval under the control of a clock.

Syntax: A set of rules dictating acceptable grammar of a computer language.

Tablet: A flat-surfaced graphic input device used with a stylus for inking and cursor movement or with a puck for digitizing.

Tape drive (magnetic): A hardware device that records and reads magnetic tape.

Tape punch: A device that punches holes in a paper tape to record coded instructions.

Tape reader: A device for reading and interpreting information stored on tape and converting it to electrical signals.

Teach pendant: The control box that an operator uses to guide a robot through the motions of its tasks.

Tensile strength: The maximum strength of a material while it is under a pulling load as it breaks.

Terminal: A peripheral device for entering data into a data processing system or for retrieving data.

Text string: A collection of characters.

Thumbwheel: A graphic input device consisting of a rotatable dial that controls the movement of a line across the display surface, horizontally or vertically.

Timesharing: The sharing of one central processor by several terminals.

Tolerance: An allowable variation in a feature of a part; the total deviation of a measurement from the desired measurement such as the distance between the minus and plus factors.

Tool center point (TCP): A given point at the tool level around which the robot is programmed for task performance.

Tool design: That division of mechanical design that specializes in the design of jigs and fixtures.

Tool life: The life expectancy of a tool, usually expressed as the number of pieces the tool is expected to make before it wears out or the number of hours of use anticipated.

Touch-sensitive display: A display device whose display surface can register physical contact.

Trackball: A graphic input device that employs a mounted rotatable ball to control the position of the cursor, used for producing coordinate data.

Translate: To transform a display item on the display surface by repositioning it to another coordinate location.

Turnkey: Any complete system of hardware, software, and service sold for a single price by one vendor.

Utility program: Any program designed to perform routine housekeeping functions in a computer system, such as copying files, making back-up tapes, removing "wasted" space on disk drives, and managing the flow of data to output devices such as plotters.

Vector: A directed line segment or a string of related numbers.

Vector plotter: A plotter that draws with ordinary ballpoint pens and ink. The pen is driven over a plotting surface by cables attached to pulleys and servomotors.

View plane: The projection plane used in three-dimensional viewing operations.

View point: The originating point of a field of view.

Viewport: A specified rectangle on the view surface within which a window's contents are displayed.

Virtual coordinate system: The result of mapping a portion of the world coordinate system to the finite limits of the device space.

Vision: The process of sensing and understanding the environment based on the light of reflectance of objects, i.e., robot vision.

Voice input device: A graphic input device that accepts and interprets vocal data.

Window clipping: The bounding of a view volume in the x and y directions by passing projectors through the corners of the window to define its sides.

Window: The specified area on the view plane containing the projections to be displayed.

Wireframe: An image of a three-dimensional object displayed as a series of line segments outlining its surface, including hidden lines.

Word: The basic unit that can be handled in a computer. A 16-bit computer handles two bytes per word.

Work center: A specific production facility or group of machines arranged for performing a specific operation or producing a given part family.

Workpiece: A part in any stage of production prior to its becoming a finished part.

Workstation: A configuration containing a display device and any associated graphic input devices.

Wraparound: The positioning of a display item so that it overlaps the border of the device space and it is displayed on the opposite side of the display surface.

Write protect: A feature that prevents the updating of a bit plane.

Yaw: Side-to-side motion at an axis.

Yon plane: The back clipping plane used in z clipping to define a finite view volume.

Zoom: To scale a display or display item so that it appears to either approach or recede from the viewer.

APPENDIX

MANUFACTURING INSIGHTS VIDEOTAPES

These videotapes are available from the:
Society of Manufacturing Engineers
One SME Drive
P.O. Box 930
Dearborn, MI 48121-0930

Phone: (313) 271-1500, ext. 418 or 419
FAX: (313) 271-2861

Automated Assembly (For use with Chapter 1.)

Automated assembly can produce consistent quality levels and shorten customer order response times. This videotape explores high speed and high volume assembly of large numbers of small units through three case studies: Tandy/Bell and Howell's Home Video Plant, Hubbell Corporation, and Remmele Engineering. *Length: 34 minutes.*

Automated Material Handling (For use with Chapter 2.)

Four case studies featuring advanced applications of automated material handling systems are presented. At General Motor's BOC complex, a highly integrated line uses one of the world's largest fleets of automated guided vehicles. Three more case studies include 3M Corp., RCA Picture Tube Division, and the US Postal Center in Buffalo, NY. *Length: 29 minutes.*

CAD/CAM (For use with Chapter 3.)

This videotape begins with a brief introduction to computer-aided design and manufacturing (CAD/CAM) benefits. Then, four case studies featuring successful CAD/CAM installations are examined including K2 Skis, Flow Systems, Inc., Toro Outdoor Products, and the Oster Division of Sunbeam Corporation. *Length: 25 minutes.*

Flexible Manufacturing Cells (For use with Chapters 4 and 8.)

This issue begins with a look at flexible manufacturing systems (FMS) and cells (FMC), illustrating how problems with original FMSs led to the development of FMCs. Four case studies featuring FMCs in action are presented. You'll visit Mazak Sales and Service, Badger Meter, Agnew Machine Company, and Lockheed-Georgia. *Length: 28 minutes.*

Robots in Assembly and Packaging *(For use with Chapter 5.)*

Five case studies help you examine the benefits of using robots for assembly and packaging. Companies visited include Comdial, Ford Motor Company, Delta Faucet, Northern Telecom, and IBM.
Length: 45 minutes.

Lasers in Manufacturing *(For use with Chapter 6.)*

See how lasers can make your machining, gaging, and marking operations easier. You'll discover how lasers convert energy into light beams that improve productivity and profitability. You then study laser applications at four locations: General Motors Saginaw Division, Cardiac Pacemakers Inc./Lilly, Allen–Bradley, and Harley-Davidson. *Length: 35 minutes.*

Lasers in Manufacturing—A New Look *(For use with Chapter 7.)*

This video program begins with an overview of lasers in manufacturing, then explores applications at Defiance Metal Products, Laser Fare Limited, Quantum Laser Corporation, and Laser Line, Inc. You'll also see why laser processing is gaining acceptance in many diverse manufacturing areas. *Length: 30 minutes.*

Programmable Controllers *(For use with Chapter 8.)*

Learn how programmable controllers process inputs to determine proper outputs, providing automatic control of manufacturing activities and process functions. You'll also visit Spectra-Physics, General Motor's BOC Orion assembly plant, Molex Inc., and Babcock and Wilcox Nuclear Equipment Division. *Length: 27 minutes.*

Flexible Manufacturing Systems *(For use with Chapter 9.)*

First, you'll discover how Mack Truck uses a traditional FMS. Next, see how a flexible system resulted in a $25 million cost avoidance in the production of B-1 bomber parts at Vought Aerospace. And last, a blend of standard and special equipment for axle housings increases productivity for an off-road vehicle manufacturer. *Length: 50 minutes.*

Personal Computers in Manufacturing
(For use with Chapter 10.)

The flexibility and compatibility of today's PC systems make various applications possible without costly add-ons or upgrades. You'll examine five in-depth case studies of personal computers at work. Also covered are integration with mini-computers and mainframes and closed-loop integration with other manufacturing systems.
Length: 28 minutes.,

Other Related Videotapes

CAD/CAM Networking

This program explores developments, applications, and trends in CAD/CAM networking. You'll see CAD/CAM networking in action at Boeing Military Airplanes, Pixley-Richard's injection molding plant, and Hydroline Corporation. *Length: 36 minutes.*

Adaptive Control

This tape shows you how adaptive control is used to reduce variability and improve predictability in manufacturing. You will examine four in-depth case studies at McDonnell Douglas, Ansu Manufacturing, Illinois Gear, and FMC. Find out why increased interest in quality will result in increased interest in adaptive control. *Length: 38 minutes.*

Simulation

Three case studies illustrate how companies used simulation to watch their plans and designs in operation before committing to construction. First, Rohr Corporation achieved greater equipment utilization in the design of an autoclave loading/unloading system and production scheduling. Then you'll visit Intel Corporation and General Electric. *Length: 28 minutes.*

Robots in Welding and Painting

Your study begins with an overview of how robots are increasing productivity and improving quality in jobs that are tough and tiring for humans. Five case studies illustrate industrial robots in action. You'll also see how forthcoming improvements will have an even greater impact in this area. *Length: 35 minutes.*

Automated Inspection/Non-Destructive Testing

See automated inspection at GKS Inspection Services where they use coordinate measuring machines to conduct dimensional inspection of parts to ensure quality. Next, visit Buick Hydra-matic, Lockheed Aeronautical Systems, and Laser Technology. You'll see new testing techniques including x-ray, ultrasound, and electronic shearography. *Length: 30 minutes.*

Sensors

This videotape explores some of the benefits and potential applications of sensors. You'll see how computers and sensors are matched to provide added capability for the metalworking industry at Werth Engineering Company. Then, you'll visit Stihl, Inc. where chain saws are produced. *Length: 30 minutes.*

Cutting Tools

First, at Kaan Engineering, inserts and high-speed steel work together to make a pump part. Next, the Bedford Gear Division of Joy Manufacturing blanks and finishes small lots of various gear sizes. Third, at Cummins Engine, cutting tools are used in facing and contouring a crankshaft vibration damper. *Length: 28 minutes.*

Machine Vision

You begin with a thorough introduction to machine vision, including various types such as binary, gray scale, and correlation. Then, see how machine vision is being used today at General Motor's BOC Orion plant, Zapata Industries, the Kearfott Division of Singer, and Seaboard Lemon Association. *Length: 50 minutes.*

Composites in Manufacturing

Your study includes a discussion on the properties of various composites, including polymer, metal, and ceramic matrix. You'll also see applications of composite technology at Glastrusions, Century Plastics, and Structural Composites Industries. *Length: 31 minutes.*

MANUFACTURING INSIGHTS VIDEO SERIES
Order Form (0979D)

NOW!

Complete your instructional program with these quality videotapes from the Society of Manufacturing Engineers. Each tape conforms directly to a chapter in the text and supports the educational process by providing a visual explanation of the featured concept and its application.

You won't want to miss this valuable opportunity—order your videos TODAY! (Quantity Discounts Available)

☐ Yes! I need the information contained in these **Manufacturing Insights** videotapes. Please send me the title(s) indicated below. (Prices include UPS shipping and handling.)

Qty.	Order No.	Tape	Price Each	*Quantity Discount	Total
	CAD, CAM, CIM				
_____	VT252-0979D	CAD/CAM (VHS)	$200.00	$160.00	_____
_____	VT252U-0979D	CAD/CAM (U-Matic)	200.00	160.00	_____
_____	VT282-0979D	CAD/CAM Networking (VHS)	200.00	160.00	_____
_____	VT282U-0979D	CAD/CAM Networking (U-Matic)	200.00	160.00	_____
_____	VT250-0979D	Personal Computers in Manufacturing (VHS)	200.00	160.00	_____
_____	VT250U-0979D	Personal Computers in Manufacturing (U-Matic)	200.00	160.00	_____
_____	VT253-0979D	Simulation (VHS)	200.00	160.00	_____
_____	VT253U-0979D	Simulation (U-Matic)	200.00	160.00	_____
	Metalworking and Control				
_____	VT254-0979D	Programmable Controllers (VHS)	200.00	160.00	_____
_____	VT254U-0979D	Programmable Controllers (U-Matic)	200.00	160.00	_____
_____	VT239-0979D	Adaptive Control (VHS)	200.00	160.00	_____
_____	VT239U-0979D	Adaptive Control (U-Matic)	200.00	160.00	_____
_____	VT249-0979D	Cutting Tools (VHS)	200.00	160.00	_____
_____	VT249U-0979D	Cutting Tools (U-Matic)	200.00	160.00	_____
_____	VT256-0979D	Flexible Manufacturing Cells (VHS)	200.00	160.00	_____
_____	VT256U-0979D	Flexible Manufacturing Cells (U-Matic)	200.00	160.00	_____
_____	VT237-0979D	Flexible Manufacturing Systems (VHS)	200.00	160.00	_____
_____	VT237U-0979D	Flexible Manufacturing Systems (U-Matic)	200.00	160.00	_____
	Composites				
_____	VT248-0979D	Composites in Manufacturing (VHS)	200.00	160.00	_____
_____	VT248U-0979D	Composites in Manufacturing (U-Matic)	200.00	160.00	_____
	Automation				
_____	VT251-0979D	Automated Material Handling (VHS)	200.00	160.00	_____
_____	VT251U-0979D	Automated Material Handling (U-Matic)	200.00	160.00	_____
_____	VT255-0979D	Automated Assembly (VHS)	200.00	160.00	_____
_____	VT255U-0979D	Automated Assembly (U-Matic)	200.00	160.00	_____
_____	VT281-0979D	Automated Inspection/Non-Destructive Testing (VHS)	200.00	160.00	_____
_____	VT281U-0979D	Automated Inspection/Non-Destructive Testing (U-Matic)	200.00	160.00	_____
	Vision, Lasers, Sensors				
_____	VT240-0979D	Machine Vision (VHS)	200.00	160.00	_____
_____	VT240U-0979D	Machine Vision (U-Matic)	200.00	160.00	_____
_____	VT242-0979D	Lasers In Manufacturing (VHS)	200.00	160.00	_____
_____	VT242U-0979D	Lasers In Manufacturing (U-Matic)	200.00	160.00	_____
_____	VT280-0979D	Lasers In Manufacturing—A New Look (VHS)	200.00	160.00	_____
_____	VT280U-0979D	Lasers In Manufacturing—A New Look (U-Matic)	200.00	160.00	_____
_____	VT257-0979D	Sensors (VHS)	200.00	160.00	_____
_____	VT257U-0979D	Sensors (U-Matic)	200.00	160.00	_____
	Robotics				
_____	VT238-0979D	Robots in Assembly and Packaging (VHS)	200.00	160.00	_____
_____	VT238U-0979D	Robots in Assembly and Packaging (U-Matic)	200.00	160.00	_____
_____	VT241-0979D	Robots in Welding and Painting (VHS)	200.00	160.00	_____
_____	VT241U-0979D	Robots in Welding and Painting (U-Matic)	$200.00	$160.00	_____

The following tapes are also available from SME but were not used in the development of the text.

Qty.	Order No.	Tape	Price Each	*Quantity Discount	Total
_____	VT290-0979D	CAD/CAM Workstations (VHS)	$200.00	$160.00	_____
_____	VT290U-0979D	CAD/CAM Workstations (U-Matic)	$200.00	$160.00	_____
_____	VT286-0979D	Simultaneous Engineering (VHS)	$200.00	$160.00	_____
_____	VT286U-0979D	Simultaneous Engineering (U-Matic)	$200.00	$160.00	_____
_____	VT284-0979D	Implementing Just-In-Time (VHS)	$200.00	$160.00	_____
_____	VT284U-0979D	Implementing Just-In-Time (U-Matic)	$200.00	$160.00	_____
_____	VT287-0979D	Factory Data Collection (VHS)	$200.00	$160.00	_____
_____	VT287U-0979D	Factory Data Collection (U-Matic)	$200.00	$160.00	_____
_____	VT291-0979D	Advances in CNC (VHS)	$200.00	$160.00	_____
_____	VT291U-0979D	Advances in CNC (U-Matic)	$200.00	$160.00	_____
_____	VT288-0979D	Tooling for Plastics and Composites (VHS)	$200.00	$160.00	_____
_____	VT288U-0979D	Tooling for Plastics and Composites (U-Matic)	$200.00	$160.00	_____
_____	VT285-0979D	Machine Vision—A New Look (VHS)	$200.00	$160.00	_____
_____	VT285U-0979D	Machine Vision—A New Look (U-Matic)	$200.00	$160.00	_____
_____	VT289-0979D	Robotics in Circuit Board Assembly (VHS)	$200.00	$160.00	_____
_____	VT289U-0979D	Robotics in Circuit Board Assembly (U-Matic)	$200.00	$160.00	_____
_____	VT283-0979D	Robotics in Surface Preparation (VHS)	$200.00	$160.00	_____
_____	VT283U-0979D	Robotics in Surface Preparation (U-Matic)	$200.00	$160.00	_____

Total Enclosed $ _____

*Quantity Discounts Available! Order six or more of these informative videotapes and save 20%! Each tape will cost you only $160.00. Order yours and save today!

Your Choice of 3 Options:

☐ Payment Enclosed
☐ Purchase Order Enclosed
☐ Credit Card (Circle One)

MasterCard   American Express   VISA

Account Number _____   Expiration Date _____

Signature _____

Ship to: (Please Print or Type)

Name _____

Title _____

School/Company _____

Address _____

City _____

State/Zip _____

Phone ( ) _____
(In case there's a question about your order.)

Mail order to: Delmar Publishers Inc.
2 Computer Drive, West
Box 15015
Albany, New York 12212-5015

**Order by phone with your credit card.
Call 1-800-347-7707.**

Note: Your order must be accompanied by full payment, purchase order, or credit card information. Prices subject to change without notice. Sorry, videotapes are not available for rent. All videotape sales are final.

Preview Policy: To discuss preview arrangements, please contact Kim Harris at 1-800-347-7707.

INDEX